QUALITY CONTROL AND ASSURANCE IN ADVANCED SURFACE ENGINEERING

QUALITY CONTROL AND ASSURANCE IN ADVANCED SURFACE ENGINEERING

Edited by
K. N. STRAFFORD
and
C. SUBRAMANIAN

CRC Press
Taylor & Francis Group
Boca Raton London New York

CRC Press is an imprint of the
Taylor & Francis Group, an **informa** business

Book 678

CRC Press
Taylor & Francis Group
6000 Broken Sound Parkway NW, Suite 300
Boca Raton, FL 33487-2742

ISBN-13: 978-1-86125-037-7 (hbk)

Publisher's Note

The publisher has gone to great lengths to ensure the quality of this reprint but points out that some imperfections in the original copies may be apparent.

Typeset in the UK by
Fakenham Photosetting

Contents

FOREWORD

All engineering components are required to interface with the environments within which they operate. Thus, for a tool or die used in the machining or shaping of a metal alloy or plastic, interaction with these types of workpiece leads to wear or even corrosion problems.

Whilst adequate bulk properties of the tool, viz. modulus and strength to resist cutting forces etc. are mandatory, particularly critical are its surface properties necessary to impart resistance to wear, especially adhesive wear. More generally, all components must exhibit appropriate compatibility with their working environment, a condition never evidently fully realised in that wear remains a major cause of engineering failures.

The modification of bulk property at the surface of a working component – *surface engineering* – by heat treatment and/or the use of a coating has long been recognised as the most efficient way to overcome, or at least alleviate, these problems.

In the field of electronics, enormous progress in the creation and synthesis of coatings, to generate e.g. semiconductor devices, has been possible through the development and use of processes such as chemical vapour deposition (CVD) and, more recently, in the last decade or so, physical vapour deposition (PVD) in its many variants. The powerful potential of these techniques to create hard tribological coatings has also been recognised and exploited, and tools and dies coated with titanium nitride have, in general, proved to be effective with significant enhancement of manufacturing efficiency. However, there is enormous scope in the possible use of other compounds generic to TiN such as TiAlN and TiAlVN etc., as well as carbides and oxides, and indeed nitrides of other metals. PVD technology, with its inherent flexibility, offers particular potential in the creation of *designed* multilayer tribological coating systems.

A key to the successful exploitation, especially commercial, of such coating systems will, however, lie in the generation and acceptance of appropriate quality assurance and quality control (QA and QC) standards. Currently, there is a proliferation of papers in the research literature extolling the benefits of many novel coatings, and coating systems. There is an urgent need to characterise, and fully understand and interpret the reported enhanced performance of such coatings, and incorporate these observations into agreed quality standards. The absence of such standards in the field of the so-called advanced surface coatings is in marked contrast to the many and still-evolving standards with other more traditional coatings.

The present book addresses comprehensively quality issues in advanced surface engineering practices. The major part of the text originates from papers read at the

Workshop on *Quality Assurance of Advanced Coatings* which was incorporated into the Second International Conference on Surface Engineering, held in Adelaide, South Australia, in 1994.

K. N. Strafford
C. Subramanian
Ian Wark Research Institute
The University of South Australia
Adelaide, South Australia, 5095. Australia

INTRODUCTION: QUALITY IN SURFACE ENGINEERING PRACTICES

K. N. STRAFFORD* and C. SUBRAMANIAN
Ian Wark Research Institute, The University of South Australia, SA 5095.

Surface engineering (SE) process technologies, to create surfaces on working engineering artefacts with enhanced resistance to wear and corrosion, have advanced significantly in recent years. It is now becoming possible to create surfaces and surface coatings from an almost infinite range of materials, to design – in specific properties – to faithfully fulfil the optimised desired engineering purpose or function. Coatings can be produced of novel composition and microstructure, in multilayer format, as appropriate and deemed necessary within the design audit, by a variety of advanced physical and chemical processes, including combination technologies. However in order for such coatings/treatments to become accepted for use in, for example, manufacturing industry, there is a need to establish, evaluate and implement appropriate and universally accepted *quality assurance and control procedures.*

Such rapid advances in process capability have, in fact, created attendant problems in the areas of quality assurance and control. The users of advanced tribological coatings such as TiN are concerned with the variability in the performance of coated components/tools in real applications. For example, although coated high-speed steel tools have typically outperformed uncoated ones by over five times, the performance of some coated tools has proved to be worse than the uncoated entities apparently due to poor coating adherence.[1] This aspect of failure could become a major problem in situations where loose debris cannot be tolerated, e.g. total hip prothesis replacement[2] or closed bearings.

Again, Hatschek[3] has reported scatter in the performance of TiN coated drills, where also poor coating adhesion was suggested to be the principal factor underlying irregular behaviour. Substrate surface roughness has also been shown to be an important factor in the coating lifetime and the approach to failure with drills.[4] More recently Doyle *et al.*,[5] in considering the anticipated successful development of alloy nitride coatings such as TiCN and TiAlN, emphasised that claims of improved performance may not always be realised in the commercial environment because of the inherent variability of PVD coating technology.

*Present address: The Institute of Materials, London SW1Y 5DB, UK.

Quality assurance must be seen as a central element in any surface engineering scenario – the defined production of a given coating or surface treatment, requiring the choice of suitable process technology or combination of technologies, with particular measurable properties and characteristics which, in turn, faithfully impart the desired performance. This matter has been emphasised by a number of research workers in the SE field – for example Weiß,[6] and Strafford and co-workers.[7–9]

Presently there are very few recognised quality standards available for the manufacturer to specify to the coating supplier, to ensure reproducibility and reliability in function for the customer or user. This situation is in contrast to the many often very specific and exacting standards which are available for more traditional surface engineering processes[10] such as galvanising, chromium plating and nickel plating which have been in existence for many years[11] and, furthermore, have been refined and made progressively more role-specific on the basis of experience.

There are numerous standards available in the literature from various national organisations (British Standards (U.K.), ASTM (U.S.A.), DIN (Germany) and AS (Australia)) relating to coating quality, in terms of appearance, thickness, adhesion, porosity, structure, hardness, corrosion behaviour, wear performance and electrical and other properties. It has taken many years for such standards to evolve and be accepted by the coating community. Initially, specifications were written for depositing a particular coating and progressively standards were produced for specific applications.[12,13]

For example, the first Australian Standard for chromium plating was issued in 1947, which was after many years of commercial application of such coatings. It was revised and improved in 1961, 1966, 1973 and 1982. These standards would further be refined incorporating the experiences and understanding gained in the future.

In advanced surface engineering – for example in the area of tribological coatings for tools and dies, coatings such as TiN and its ternary and quaternary derivatives, TiAlN, TiAlVN etc. – no comprehensive quality standards have been published. In fact, however, the whole basis for the establishment of suitable property/performance parameters, on which suitable quality could be confidently based, is not well established. Thus the precise nature and significance of the link between characteristics and properties such as composition, structure, morphology, surface roughness, hardness, internal stress level, adherence etc., and performance (corrosion and wear resistance), is not well characterised. If this link could be better established, and its basis understood, ideally on a fundamental basis (i.e. in accordance with well-accepted and proven physical metallurgy principles), then, ultimately, this linkage could be used through an expert system approach to control, in the most effective manner, the coating/surface treatment process parameters in order to obtain reproducible optimum performance demanded by the manufacturer and especially the user of the coated/surface treated component.[7,9]

The present book addresses some of these problems – both philosophical and practical – and seeks to provide a timely authoritative set of chapters establishing the need for, and the role and status of QA/QC matters in the field of the advanced surface engineering technologies.

REFERENCES

1. M. McPhee: 'Studies of the Performance of Coated High Speed Cutting Tools', *PhD Thesis*, University of South Australia, October, 1995.
2. L. P. Ward, K. N. Strafford, T. P. Wilks, C. Subramanian and T. V. Parry: 'The Sliding Wear Behaviour of Selected Metal Nitride Coatings for Possible Use in the Construction of Metal–Metal Joint Replacements', *Proc. Second Int. Conf. on Surface Engineering*, Adelaide, Australia, March 1994.
3. R. L. Hatschek: *American Machinist*, Special Report 752, 1983, 192.
4. C. Subramanian, K. N. Strafford, T. P. Wilks, L. P. Ward and W. McMillan: 'Influence of Substrate Roughness on the Scratch Adhesion of Titanium Nitride Coatings', *Surface Coatings and Technology*, 1993, **62**, 529.
5. E. D. Doyle, J. H. Hamlyn-Jones and T. Phan: 'The Next Generation of PVD Coatings for Cutting Tools', *Tools '94 Conf.*, Sydney, Australia.
6. H. Weiß: 'Characterisation and Quality Assurance of Surface and Surface Coatings', Ch. 14 in *Surface Engineering: Processes and Applications*, K. N. Strafford, R. St. C. Smart, I. Sare and C. Subramanian eds, Technomic Publications, Pa, USA, 1995.
7. K. N. Strafford, C. Subramanian and T. P. Wilks: 'Relationship and Linkages between Coatings Performance in Manufacturing Operations and Properties and Characteristics: The Origin of Quality Issues', *J. Materials and Processing Technology*, 1993, **38**, 431.
8. K. N. Strafford, C. Subramanian and T. P. Wilks: 'Characterisation and Quality Assurance of Advanced Coatings', *Surface Engineering*, Vol. II, *Engineering Applications*. P. K. Datta and J. S. Gray, Royal Society of Chemists, London, 1993, ch. 2.5.1, 259.
9. K. N. Strafford: 'Tribological Properties of Coatings – Expectations, Performance and the Design Dilemma', Paper presented at First Australian–U.S.A. Workshop on Critical Issues in High Performance Wear Resistant Films, Wollongong, N.S.W., Australia, February, 1995. *Surface Coatings and Technology*, in press August 1995.
10. K. N. Strafford, P. K. Datta and J. S. Gray: 'Applications and Performance of Coatings: Some Perspectives and Prospects', *Surface Engineering Practice: Processes, Fundamentals and Application in Wear*, Idem. Ed., Ellis Horwood, Chichester, UK, 1990, ch. 3.1.1.
11. D. R. Gabe: 'Metallic Coatings for Protection', Ch. 4 in *Coatings and Surface Treatments for Corrosion and Wear Resistance*, K. N. Strafford, P. K. Datta and C. G. Googan, eds, Ellis Horwood, Chichester, UK, 1984.
12. V. E. Carter: *Corrosion Testing for Metal Finishing*. Butterworths, London, 1982.
13. K. A. Chandler and D. A. Bayliss: *Corrosion Protection of Steel Structures*, Elsevier, Lausanne, 1985.

ORIGINS AND ESTABLISHMENT OF QUALITY ASSURANCE WITH ADVANCED SURFACE COATINGS

K. N. STRAFFORD*

Surface Engineering Research Group, Department of Metallurgy, The University of South Australia SA 5095, Australia

ABSTRACT

Generally, the traditional, well-established industrial practices of surface engineering have developed over many years and there is often an historical dimension which is punctuated with particular well-recognised, approved and published quality standards, each reflecting critical stages in these developments, and the continuing drive to achieve true *fitness for purpose*. In contrast, advanced surface engineering techniques, perhaps epitomised by the physical vapour deposition (PVD) ion plating processes, have developed very rapidly over the last decade or so. As a result, there is still much yet to be researched and learnt about advanced coatings so-produced (such as TiN, ZrN, TiAlN etc.), not only in terms of their basic structure/property relationships, but especially in relation to their reliable, quantifiable performance – for example, in tribological application areas such as tooling and dies. Within a brief review of some of the factors involved in the design of such coating systems, significant basic properties and characteristics of hard coatings materials are identified, together with corresponding requirements in their use as coatings. The limited information published in the open literature describing the established correlations between certain of these properties/characteristics of coatings and their tribological performance is reviewed, information of the type which could be used to begin to establish quality control and assurance parameters. Finally, the need for a national and international approach to this critical topic area, including a focussed research and development effort, is briefly considered.

1. INTRODUCTION

In recent years there has been a continuing and rapid development of advanced surface engineering practices[1,2] for the optimised enhancement of properties such as corrosion and wear resistance of engineering artefacts. It is now possible to produce coatings of novel composition and microstructure in multilayer format, as appro-

* Present address: The Institute of Materials, London SW1Y 5DB, UK.

priate to the perceived or actual design audit, by a variety of advanced physical and chemical processes, including combination technologies. The extraordinarily rapid pace of advances and developments in the field needs to be recognised, especially over the past decade or so, as a result of which often there is not a deep understanding of why particular coatings or coatings systems perform well (or badly); that is, there is often an insufficient knowledge of basic structure/property relationships in the coatings materials themselves, and especially in the way such coatings, when in intimate contact with a chosen substrate of 'correct' design and properties, e.g. of sufficient rigidity, form a coating system to engender improved (or otherwise) wear performance.

For example, there is poor correlation between the hot hardnesses of the Group IV metal nitrides – TiN, ZrN and HfN – and the cutting performance of appropriate tools, coated with these compounds.[3] Again, the exceptional hardness which can apparently be achieved with compounds generic to TiN, e.g. TiAlN, TiAlVN, by judicious control, *inter alia*, of PVD processing parameters through (presumably) associated variations in stoichiometry and microstructure, have not unequivocally and reliably been translated into greatly enhanced tool life. Even with a better understanding of the basic structure/properties relationships with such novel compounds, their performance within a coating system will be enormously influenced by the adhesion of the coating to the substrate, itself a function of many variables, where the relative differences in the nature of the coating and the substrate, e.g. Young's Modulus, Poisson's ration, coefficient of thermal expansion, thermal conductivity etc. are obviously critical.[4]

Generally the problem is one of *compatibility* between the coating and substrate. Strafford *et al.*[5] have pointed out that, in principle, no single coating can satisfy the totally different compatibility demands at the workpiece/coating interface on the one hand, and those of the coating/substrate (tool) interface on the other. Surface conditions, cleanliness and finish are also of major significance, factors which it is believed all too often are ignored or, at least, not characterised and standardised as prerequisites for repeated coating success and achieving reliable, desired performance.

From the perspective of a user of coatings such details and subtleties must be seen as a nicety. The unreliable performance which has been reported with even 'simple' coatings such as TiN, is however, a matter of concern. The real value of coated tools for the manufacturer concerned with optimisation and efficiency, particularly within the CNC machine tool scenario, begins to be questioned. Also, the enthusiasm, drive and finance needed to research and develop advanced coating systems involving hard materials of novel composition and structure, possibly in multilayer format, begins to wane.

Quality assurance must be seen as a central element in any surface engineering scenario – the defined production of a given coating or surface treatment requiring the choice of a suitable process technology, or combination of technology, with particular measurable characteristics, which, in turn, faithfully impart the desired performance.

In attempting to improve the quality and reliability in the performance of coated tools there is however a need to recognise the urgent requirement to *establish* appropriate quality control procedures and assurance standards. The absence of such parameters and recommended procedures becomes very evident for advanced coatings, in contrast to the well-established standards for the more traditional corrosion-resistant coatings, e.g. chromium, zinc, nickel etc. What is striking here is the way in which such standards have become modified, adapted and improved over the years as the properties and required performances of the coatings have become more focussed and specific. There is, then, an evolutionary factor at work in the establishment of quality performance standards.

The object of the present paper is to briefly review the various factors involved in the design of a successful tribological coating – *fitness for purpose*; examine the reported correlation between tribological performance and the characteristics and properties of various hard materials, so allowing the identification of possible QA and QC parameters; finally, suggestions are made as to possible ways in which national and international standards for advanced coatings may be agreed and drawn up, to the obvious benefit of the tool and die user.

2. THE DESIGN AND PERFORMANCE OF TRIBOLOGICAL COATING SYSTEMS: AN OVERVIEW

2.1 The Coating System: Key Features

Initially it is pertinent to consider some of the basic features of a working coating system – for wear (or corrosion) resistance – as discussed by Strafford *et al.*[5] Generalised features of such a system are illustrated in Fig. 1. In simple terms, any coating system – comprising together the key entities of coating and substrate – is required to relate or respond appropriately to an operating envelope, described as either the working environment or counterface. The former terminology usually is employed in the context of protective coatings and corrosion, and the latter with reference to manufacturing processes. Although the coating and substrate together constitute the coating system it is most important to recognise the marked, distinct, interface which separates these entities – Interface 2 in Fig. 1. Interaction between the coating and the substrate across this interface may lead to performance failure, constituted by excessive interdiffusion during fabrication (e.g. in the production of a diffusion coating via e.g. chromising or aluminising) and/or during service especially at the elevated temperatures experienced, for example, in the operation of gas turbine blades, or in fast cutting or forming operations. Loss of coating adhesion may occur for a variety of reasons: the coating itself may also suffer loss of cohesion.

More familiar and recognisable are the interactive damage modes or processes which occur across Interface 1. The damage modes here are, in general terms, described as corrosion or wear, the latter taking place either at ambient temperatures

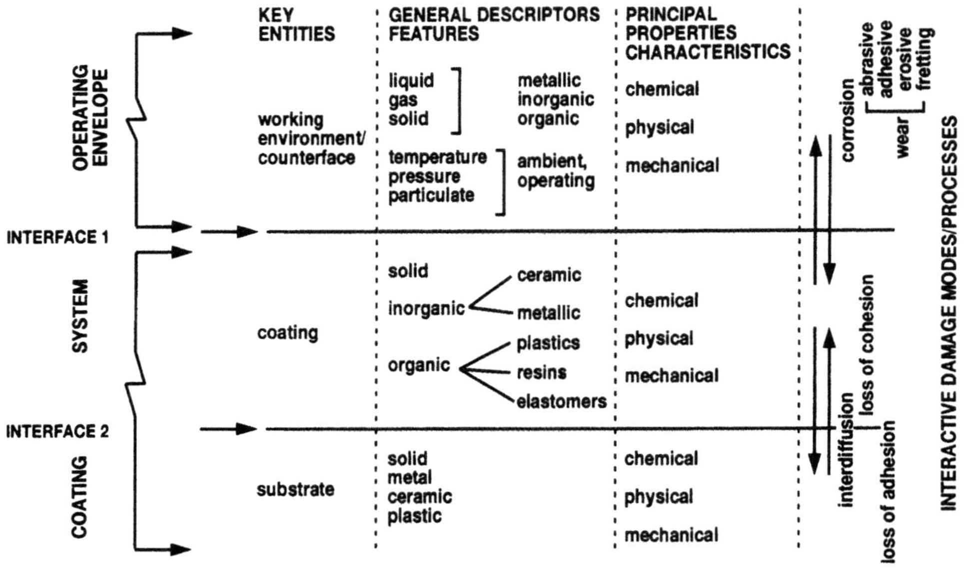

Fig. 1. Generalised features of a working coating system.[5]

(often in aqueous environments – 'wet' corrosion) or at elevated temperatures (often referred to as 'dry' corrosion). Wear processes are many and complex but can be satisfactorily classified[6,7] as abrasive, adhesive, fatigue or corrosive, depending on the nature of the environment/counterface: that is, its descriptors or features, and its principal properties or characteristics (as set out in Fig. 1), and the coating itself, also similarly identified.

Generally, it is to be noted that the fundamental requirement is *compatibility*, broadly defined. Thus total incompatibility between the substances comprising the coating and the environment/counterface is required in an ideal working system. On the other hand good compatibility between a coating and substrate is necessary to be able to create a sound coating system. It is therefore clear that since these fundamentally different requirements ideally have to be met across the Interfaces 1 and 2 in Fig. 1, no single coating can simultaneously satisfy such requirements, and a degree of compromise has to be achieved in real, practical systems. Nevertheless, in general, adequate systems can be, and are, developed. However, the advent of modern advanced process technologies, perhaps even in combination, offer the prospect of greatly improved and reliable, optimised, performance – *fitness for purpose* referred to earlier – **provided adequate proven quality assurance guidelines can be agreed and established**. Herein lies the root of the problem.

2.2 Tribological Coatings: Some Design Aspects

Subramanian and Strafford[8] have recently examined the principles underlying the design of coatings for tribological applications, and it is useful to consider some of the issues raised in this paper in the context of the theme of the present article.

It is pertinent to consider three aspects: (1) the availability and choice of the various possible coatings **materials** which can be formed from combinations of specific metallic and non-metallic elements in the Periodic System; (2) the possible variations in the **structure** or **architecture** of coatings; and (3) the reported behaviour **performance** indicators, in outline, of particular designs of coating systems.

2.2.1 Coatings materials: scope

There are a large number of materials – metal carbides, nitrides, oxides and borides – which may be considered for service to resist wear (and corrosion), which are metallic, covalent or ionic in nature. Particular interest has centred on MX-type compounds, where M is a metal and Z a non metal (B, C, N, O), involving Groups III A (Al), IVA and IVB (Si, Ti, Zr, Hf), VB (V, Nb, Ta) and VIB (Cr, Mo, W) metals. Ternary $(M_1\ M_2)X$, and quaternary $(M_1\ M_2\ M_3)X$-type hard compounds, derived from various combinations of these metals, have also been considered and researched such as TiAlN and TiAlVN. Subramanian and Strafford[8] have summarised various combinations which have been used – see Tables 1 and 2.

Table 1. Various compounds used as coating materials for tribological applications (data from various sources).[8]

	IV B			V B			VI B			IIIA		IV A
	Ti	Zr	Hf	V	Nb	Ta	Cr	Mo	W	B	Al	Si
B	•	•				•						
C	•	•	•	•	•	•	•		•	•		•
N	•	•	•	•	•	•	•	•		•	•	•
O	•	•					•				•	
CN	•											
ON	•											
OC	•											
OCN	•											

Aspects of the scientific rationale behind the synthesis of such compounds have also been discussed. Modification of properties of a binary compound such as TiN may be achieved by partial substitution of the Ti sublattice by metals such as Al, V, Zr or Hf. A variation in the relative concentration of the non-metallic element, e.g. C and N, alters the valency electron concentration (VEC) in the phase, also accompanied by changes in mechanical and physical properties. In this context the effects of VEC on hardness in certain carbides, mixed carbides and carbonitrides has been established[9] (see Fig. 2), but much more systematic research is needed for other significant properties (see Section 3) such as toughness, chemical stability and wear resistance to be monitored and understood as a function of VEC, in order to begin to *predict* the performance of coatings from such basic principles. As would be

Table 2. Ternary and quaternary hard compounds used as tribological coatings (data from various sources).[8]

	Ti											
	Zr	Hf	V	Nb	Ta	Cr	Mo	W	Al	Si	AlV	AlZr
B												
C									•	•		
N	•	•	•	•	•	•			•		•	•
O									•			
CN			•						•	•	•	
ON									•			
OC									•			
OCN												

anticipated, stoichiometry in all these types of compounds greatly affects properties,[10] and in order to achieve reproducible properties and therefore behaviour in a coating material, composition must be strictly controlled. Some of the ways in which coating properties have been shown to be altered and how performance is influenced by stoichometry, as affected, in turn, by the precise conditions of deposition in the PVD process, are illustrated below in Section 3.

In summary, there are a large number of potential compounds whose basic properties are appropriate and attractive and which might fulfil useful roles as coatings for the inhibition of wear. Several of these compounds have been explored especially as bulk engineering ceramics but much research remains to be conducted to be able to successfully and economically synthesise such hard materials as coatings

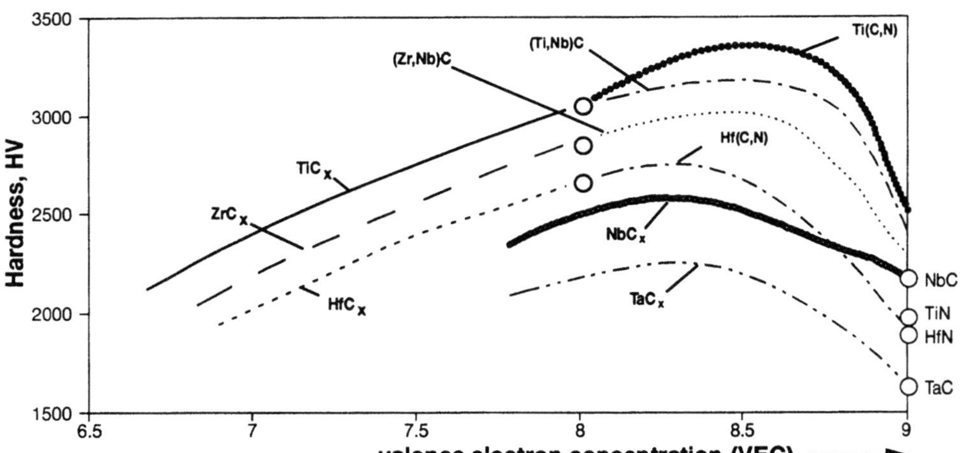

Fig. 2. Microhardness of carbides, mixed carbides or carbonitrides as a function of the valence electron concentration.[9]

with available and developed process technologies, suitable for large scale applicability.

2.2.2 Coating structure and architecture

It has been pointed out[8] that many combinations of various substrate and coating materials, i.e. coating systems, have been studied over the past 30 years or so. Subramanian and Strafford[8] have defined and illustrated a number of recognised and identifiable *structures* – sometimes called the *architecture* – of coatings: alloyed or multi component (e.g. TiAlN, TiCN); multiphase (e.g. TiC/TiN); composite (e.g. TiC/Al_2O_3); multilayer (e.g. $TiC/TiN/Al_2O_3$) and graded coatings (e.g. TiN/TiCN/TiC) – see Fig. 3.

As noted earlier, it is self evident that no single layered coating can satisfy fully the different demands at the substrate (tool)/coating interface, and the coating/workpiece interface (see Fig. 1). Particularly critical in the performance of a coating is its adhesion to the tool or die: without sufficient adhesion premature coating failure is inevitable, even with a novel or new coating material with 'ideal' intrinsic properties and characteristics which should closely satisfy the requirements of the working interface.

To improve adhesion and other mechanical and physical properties one or more intermediate layers may be incorporated into the coating system. The multilayer coatings system so produced exhibits better wear performance due to the enhanced toughening via crack deflection at the many interfaces.[8] In principle such multilayer coatings may be produced by several process technologies, singly or in combination, e.g. plasmathermal spray, electroplating, electroless plating etc. However, in the context of hard tribological coatings it is useful to review the rationale and success to date of those multi-layer systems developed via the advanced coating processes of CVD and PVD, both with recognised advantages and disadvantages.[11] As might be expected, major differences in the process details will determine (and perhaps dictate) the architecture of the coating. Since CVD processes operate at much higher temperatures than their PVD counterparts, coating interfaces become less well-defined on account of the interdiffusion which takes place during processing. Often

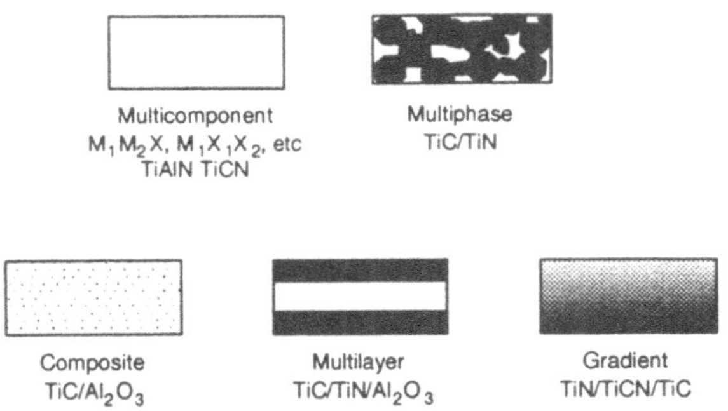

Fig. 3. Schematic illustration of the structure of different types of coatings.[8]

such interdiffusion improves adhesion if brittle intermediate phases are not formed. The higher deposition rates of PVD, especially in its magnetron variant, permit the more rapid build up of material and allow easier production of multilayer coatings with many layers. Hollek *et al.*[12,13] have thus described the design of a TiC/TiB_2 system with as many as 1000 layers, each some 5 nm thick in a coating of overall thickness of 5 μm.

Nevertheless CVD processing, dating from circa 1968, has been widely used for multilayer coating preparation. Successful commercial coatings have been developed by, for example, Sumi Tomo (Japan), by Kennametal (USA) and Sandvik (Sweden). Such coating systems frequently employ combination layers of Al_2O_3, TiN and TiCN, outer layers → inner layer, TiN and TiCN providing good substrate adhesion, Al_2O_3 wear resistance. Compositional variations may be used to give gradation in properties (see Fig. 3).

CVD and PVD technology may also be used in combination – CVD, as noted, providing for good interdiffusion and bonding adhesion of the undercoat, while PVD yields a fine grained microstructure with better toughness and wear resistance. The advantages in the application of plasma-assisted CVD or PVD in providing for

Layer Material Concept	**Preparation Method** ☐ well suited ☐ partly suited
Solid Solution Coatings	**PS, LPPS CVD, PACVD PVD (MS, IP, IB)**
Gradient Coatings	**PS,LPPS CVD, PACVD PVD (MS, IP, IB)**
Multilayers	**PVD (MS. IP, IB) CVD, PACVD**
Metastable Layer Materials	**PVD (MS, IP, IB) PACVD**
Multiphase Layer materials	**PVD (MS, IP, IB) CVD, PACVD, PS LPPS**

PS = Plasma Spraying
LPPS = Low Pressure (Vacuum) Plasma Spraying
CVD = Chemical Vapour Deposition
PVD = Physical Vapour Deposition (partly reactive)
PACVD = Plasma Activated Chemical Vapour Deposition
MS = Magnetron Sputtering
IP = Ion Plating
IB = Arc Evaporation ("Ion Bond" Method)

Fig. 4. Suitability of various coating techniques for advanced multilayer system.[14]

microstructural refinement have also been noted. Hollek[14] has discussed the several possible process routes (Fig. 4) to fabricate the various coating architectures.

Subramanian and Strafford[8] have provided a list of investigations concerning the composition, constitution and architecture of mulilayer coatings systems in the literature, and associated process technologies. The complexity of this and other similar information emphasises the difficulty in setting up agreed and recognised QA and QC proceedings for such novel advanced coating systems. Thus while the majority of total coating thicknesses in such systems was between 5 and 10 μm, the thickness of individual layers varied from a minimum of 0.003 μm to a maximum of 8 μm. Further, of the 25 or so systems cited, 17 involved two compounds, 8 three compounds while only 1 system was constituted of four compounds. The most common compound in use was TiC, a reflection on the current dominance of CVD processing.

2.2.3 Performance indicators

Finally, it is of interest to consider in this section the performance of such types of coatings, especially relative to single layer coatings. Generally the literature contains many references to the performance of particular coatings and coatings systems including those in multilayer format. In attempting to analyse these data it must be remembered that good performance in one particular tribological situation may not be upheld in another set-up. For example, the detailed property requirements of a coating for a twist drill under particular cutting conditions will be different for a lathe cutting tool. Again, die coating requirements will differ considerably for injection moulding of thermosets compared with those for metal pressing. In some instances there may be a corrosion element in the operating envelope (Fig. 1), e.g. in the moulding of PVC. Figure 5, by way of example, illustrates the potential of multicomponent or alloyed (TiAlN) and multilayer Ti(CN)/Al–O–N[15,16] coatings, while Fig. 6 exhibits the better performance of multilayer TiC/TiB_2 coatings.[12]

Clearly central in any approach to produce QA and QC standards for advanced coatings will be an agreement as to recognised limiting types of performance evaluation for various material fabrication/manufacturing operations. Here the extent to which laboratory type tests are able to simulate *in situ* full scale tool and die operations in the factory is a particularly critical area.

3. CHARACTERISTICS AND PROPERTIES OF HARD COATINGS MATERIALS AND THE CREATION OF AN OPTIMISED TRIBOLOGICAL COATINGS SYSTEM

Strafford *et al.*[5] have broadly identified key coatings' properties and requirements – listing mechanical, physical and chemical attributes – in order to develop the 'ideal' coating – see Table 3. Further, they have pointed out the need to also consider coating/substrate relationships in order to create an optimised coating system.

The need for compatibility between the coating and the substrate (crucial for

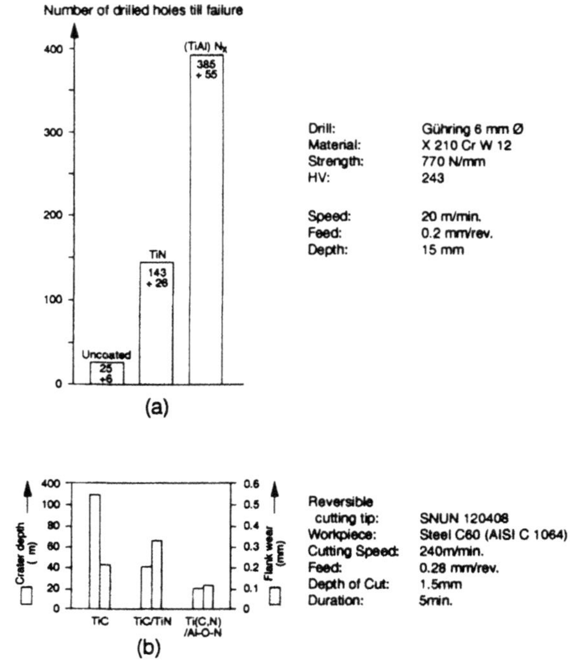

Fig. 5. (a) Drilling tests with various coatings;
(b) crater and flank wear of reversible cutting tools.[15,16]

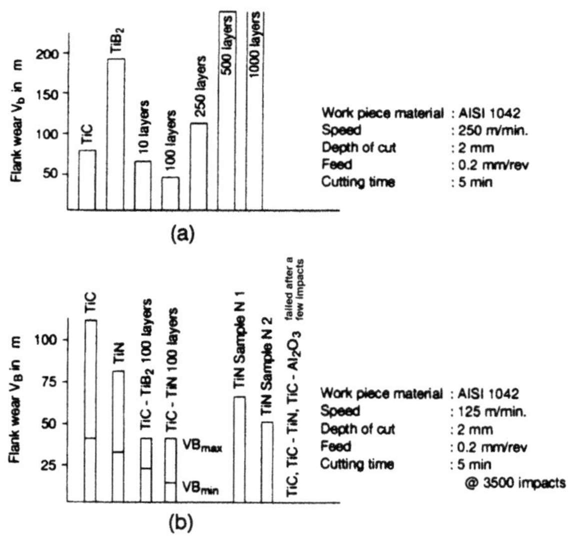

Fig. 6. Flank and crater wear of hard metal bits with TiC/TiB_2 multilayer coatings: (a) continuous cutting: (b) interrupted cutting conditions.[12]

Table 3. Key property requirements for the development of an ideal coating.

	Properties	Requirements
Mechanical	• hardness	high
	• strength (compressive and tensile)	high
	• ductility	moderate
	• toughness	moderate
	• elastic modulus	high
Physical	• thermal conductivity	high
	• coefficient of thermal expansion	low
	• melting point[1]	high
	• coefficient of friction[2]	
	• density (porosity)	low
	• crystal structure	simple, high symmetry
Chemical	• chemical compatibility[3]	good
	• composition	close to stoichiometric

[1] Relative to operating or developed service temperature.
[2] Relative to counterface-related, inter alia, to chemical compatibility, presenting either an inert surface or a surface which degrades at an acceptable rate to yield a corrosion product (commonly oxide) which functions as a lubricant.
[3] Relative to working counterface/environment.

deriving and maintaining coating adhesion) was stressed earlier in Section 2 of this paper – see Fig. 1. Thus there is a need for:

(i) low and matching coefficients of expansion:
(ii) matching thermal conductivity:
(iii) high and comparable elastic moduli;
(iv) chemical compatibility.

Thermal conductivity values will be particularly important in determining heat transfer between workpiece and tool and influence cutting temperature, and therefore performance. Under (iv), a favourable negative free energy change for coating/substrate interaction to develop a high interfaced bond strength and promote alloying should be accompanied by favourable (slow) kinetics of inter-diffusion at service temperature.

The importance of engineering the substrate to match the requirements and support of the coating is also evident – it may be appropriate to harden the substrate, for example, by ion implantation or plasma carburising/nitriding to provide matching values across Interface 1 (Fig. 1). This constitutes *duplex surface engineering* as discussed by Bell[17] and obviously introduces additional dimensions to QA and QC issues.

Figure 7 indicates the many factors which determine the quality of a coating,

Fig. 7. Interactive parameters involved in quality assurance of advanced coatings.[4]

where here 'quality' is synonymous with fitness for purpose. Quality is determined by four key interactive parameters, each with sub-set detailing.[4]

Under the heading Properties and Characteristics are listed seventeen basic and fundamental features/attributes of coatings, all of which (in relation to the discussion put forward earlier concerning the performance of a working coating system, see Fig. 1) must be considered to be of major importance. Such a list, if not absolutely comprehensive, provides for a very full description ('fingerprint') of a coating. In the real world of the coating engineer the question arises, however, as to the relative importance of items in this list, not least because of the difficulty of measuring such properties, but also, more importantly, because of the uncertainty of their real significance and value in the present relatively poor state of understanding of coatings and coatings system design and behaviour, where empiricism tends to rule. Also the interactive nature of these coatings' features needs to be recognised.

Within this listing however it is possible to immediately recognise intrinsic basic properties and characteristics which are plainly important, viz coating *hardness*, *strength*, *fracture toughness*, *ductility*, *elastic modulus*, properties which are commonly available in the literature with reference to coatings materials when in *bulk form*, often in the form of engineering ceramics: data for coatings are however scarce. For example, hardness data are confined to a few nitrides and carbides out of the many compounds of potential interest (see Tables 1 and 2), as summarised in Table 4 taken from Martin.[18] Also noteworthy in this Table are the often major differences reported between the hardness values recorded for bulk materials and their coating counterpart. The large *range* of values quoted for coatings, bracketing the single value of corresponding bulk ceramics, should also be noted. This implies

Table 4. Hardness values for hard/bulk and thin film materials.[18]

Materials	Bulk	Films
TiN	2000	1500–4000
ZrN	1500	700–3600
HfN	1600	1850–2400
VN	1500	600–2000
NbN	1400	1100–3000
TaN	1000	–
SiN	1000–2000	2000–3000
TiC	3500	2770–4150
VC	2500	1900–2850
WC	2050	1800–2800
TaC	2200	1280–2200

considerable variability in the measured hardness of coatings, as affected by other items set out in Fig. 7, both under the same heading Properties and Characteristics (e.g. composition, stoichiometry, structure morphology, microstructure) and the other principal heading Processes (to include both process type and variables). Some variability or scatter, no doubt, also arises because of the practical experimental difficulties of measuring the hardness of coatings, especially in thin film form. The particular type of apparatus/equipment used will also be important. Other key property data for coatings, e.g. fracture toughness, are not available, again partly relating to the difficulties of establishing experimental techniques. Microstructural and compositional data for thin films can be obtained thanks to the development in recent times of sophisticated surface analytical techniques as reviewed in this Conference by Smart.[19]

Summarising, it can be said that while the essential criteria-properties and characteristics of hard coatings in the development of the optimised coating systems can be recognised (including the matching requirements for suitable substrates), detailing and quantifying such criteria is a major difficulty. Property data for the many potential coatings materials are not readily available in the literature, this stemming from the problems in devising methods and equipment to enable appropriate measurements to be made. Such data that are available, e.g. hardness, are scattered and differ markedly from the corresponding values published for bulk materials. However, a more fundamental question arises as to the relevance and significance of such data that are available in the context of the tribological performance of the derived coatings systems and the drive to establish appropriate QA and QC operating parameters.

4. THE CREATION OF QUALITY STANDARDS: COATINGS' PROPERTIES AND PERFORMANCE RELATIONSHIPS

In addressing the need to establish quality procedures and practices in the area of advanced coatings, in order to be able to optimise and guarantee performance for the user, it is useful to focus initially on the ideas within and information set out in Figs 1 and 7. For a coating to perform it has to meet stringent requirements both at the working Interface 1 and the coating/substrate Interface 2 (see Fig. 1). Absolutely critical to functioning is the initial achievement of appropriate *adhesion*, and its subsequent *maintenance*, between the coating and substrate: if adhesion is satisfactory then attention turns to the problems of containing the various tool wear mechanisms, essentially at Interface 1.

What factors influence adhesion? It is suggested that this problem has to be addressed at two levels. In the first place, throughout the field of coatings technology – organic, inorganic, metallic – it is well recognised that central to good adhesion is the requirement for a *suitable surface* of the artefact to be coated. Here 'suitable' covers obvious aspects such as surface cleanliness and finish, and also more subtle factors such as hardness and internal (residual) stress levels in the surface. Any change in hardness or internal stress occurring during processing, which may be quite likely in the realms of CVD with its relatively high process deposition temperatures, may be important. Consistent surface conditions and finish are essential for reproducibility in the coating, and in any QA procedures there will be a prime need to define specific requirements on the basis of established agreed experience and experiment.

The literature contains several references to the significance of substrate surface finish on coating adherence in the area of advanced hard coatings. A recent paper[20] by Subramanian *et al.* provides a convenient summary. The whole question of surface roughness and finish is a continuing critical area in manufacture. The costs of machining/finishing to unnecessarily high standards should be avoided. In the context of hard coatings the question arises as to the *optimum* surface finish for coating adherence and *adequate performance.*

Several authors[21] have reported scratch adhesion critical loads for TiN coatings deposited on several substrate materials prepared to different roughness R_a values ranging from 0.04–4 μm. Such loads varied considerably from minimum values of some 11–15 N, to maximum values in the range ~50–70 N. No simple correlation with R_a value is evident from these data, and it becomes clear that, in addition to surface roughness, other parameters have a pronounced effect on film-to-substrate adhesion including substrate material, substrate hardness and coating thickness (in the range ~2–8 μm), parameters also cited in Fig. 7 as critical coatings' properties and characteristics.

Studies in these laboratories[20] have confirmed the marked dependence of critical loads for delamination on substrate surface roughness, these decreasing almost linearly as illustrated in Fig. 8. With the thin (3 μm) films tested it was also observed that the original surface finish of the substrate were faithfully reproduced in the

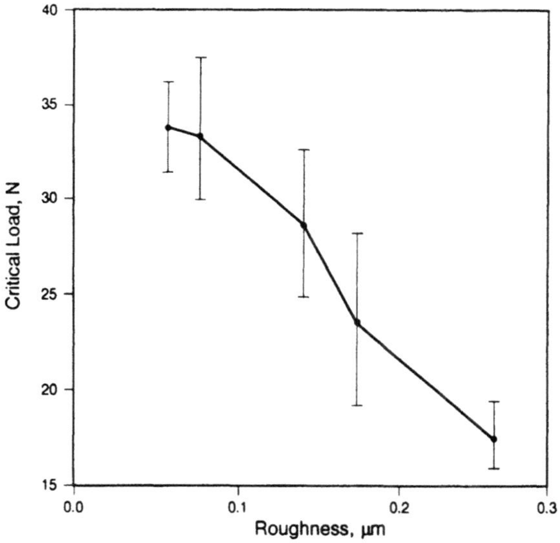

Fig. 8. The critical load to failure vs. the roughness of the steel substrates.

coating: this 'inheritance' of substrate characteristics by the coating is an important observation. The transfer of internal stress of a substrate to a coating grown on that substrate is also well established in the field of electrodeposition.

Morphological examination of delaminated surfaces indicated[20] that the failure modes of the TiN coatings on high speed steel plates could be attributed to spallation and tensile cracking, independent of surface roughness. There were significant differences, in detail, in the observed loads for detachment when measured in the transverse and parallel directions to grinding.

The second critical area for coating adhesion at Interface 2 (Fig. 7) concerns the *maintenance of adhesion*, i.e. factors which influence adhesion during the tribological service of the artefact. Here the focus shifts to a consideration of key properties and characteristics of the coating and the substrate – factors which will determine the level of mutual compatibility between the coating and substrate, as noted earlier in Section 3 above – matching (or otherwise) expansion coefficients, thermal conductivities and elastic moduli, within an overall demand for chemical compatibility.

The considerations – critical aspects of coating systems design – lie outside the control of the coater. It is, however, very much the concern of the engineer involved in a coatings R and D programme, and especially relevant to the drawing up of coatings specifications.

As noted at the beginning of this article, the coating property requirements at the coating/workpiece Interface 1 in Fig. 1 are different from those at Interface 2. Essential criteria include low frictional characteristics, relative to the workpiece, and a group of suitable mechanical properties (Fig. 7) which will permit the optimum distribution and bearing of contact stresses through the coating thickness. In detail, the precise tool function (whether involved in drilling, turning, drawing etc.) reflecting different loadings (essentially static or dynamic) will be important, as will the workpiece material.

As noted earlier in Section 3, comprehensive actual experimental data for coatings' properties are not generally available, which severely limits analysis and mathematical modelling, and therefore a systematic approach to coatings design. More disconcerting, however, in the context of the theme of this paper, is the lack of published information detailing coating property/characteristics, and coated tool performance. It follows that the rigorous establishment of QA and QC parameters for (optimised) performance is, at this stage of accumulated knowledge, fraught with difficulty.

It has been pointed out[22] that many of the properties which the coatings' researcher measures (composition, thickness, adhesion, friction coefficient, wear rate) are different from those required by the design engineer (Young's Modulus, Poisson's ratio, coefficient of thermal expansion and conductivity, specific heat, stress/temperature behaviour). However, of more concern, is the variability reported in even a limited range of coatings properties, as influenced, *inter alia*, by process variables. Coating thickness, hardness, adhesion, friction, wear, composition and residual stress have all been suggested as significant parameters which can be relatively easily and routinely measured for quality control purposes.[5]

In the literature authors have noted (for TiN) relationships between, for example, hardness, critical load for delamination detachment, internal stress and substrate bias voltage (Fig. 9), and between hardness and colour (indirectly a measure of composition, stoichiometry) and hydrogen and nitrogen gas pressures in the PVD process (Fig. 10).

Furthermore, limited experimental data are available which indicate important linkages between such measures and standardisable properties, and performance,

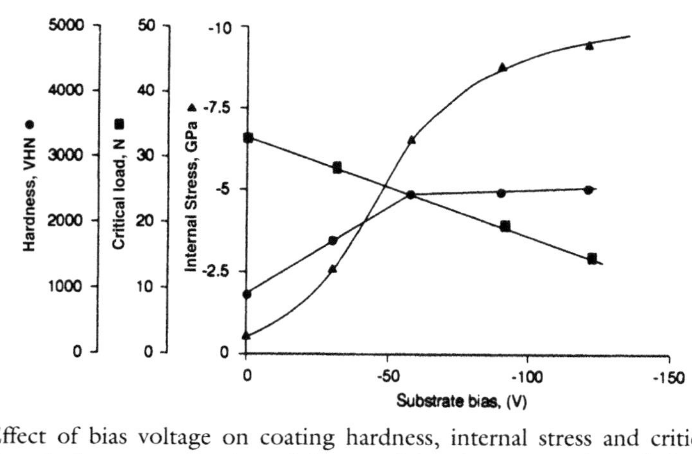

Fig. 9. Effect of bias voltage on coating hardness, internal stress and critical load for coatings deposited onto austentic stainless steel by the sputter ion plating process.[10] Courtesy of Elsevier.

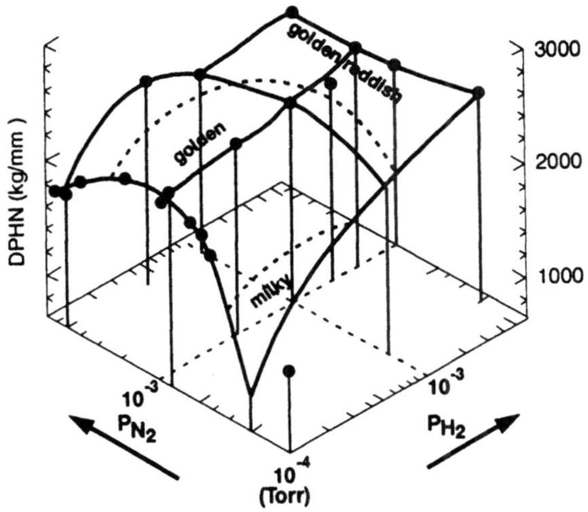

Fig. 10. The relations between microhardness, colour and hydrogen and nitrogen partial pressures for Ti–N films.[24] Courtesy of Elsevier.

e.g. between critical load and coating thickness (Fig. 11), between tool life and indent adhesion (Fig. 12), and between flank wear resistance and coating thickness (Fig. 13). Other results indicate important linkages between such measurable and standardisable properties, and performance, e.g. between critical load and coating thickness (Fig. 11), between tool life and indent adhesion (Fig. 12), and between flank wear resistance and coating thickness (Fig. 13). Further results, obtained in these laboratories, are reported elsewhere in this Conference.[28]

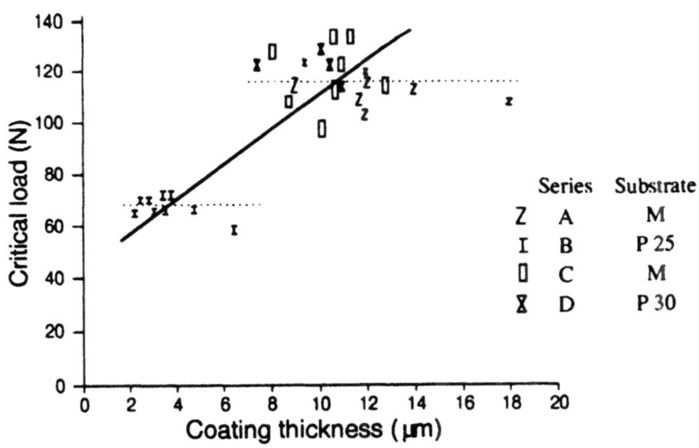

Fig. 11. The critical loads for coating detachment in scratch testing as a function of the coating thickness for various coatings.[25] Courtesy of Elsevier.

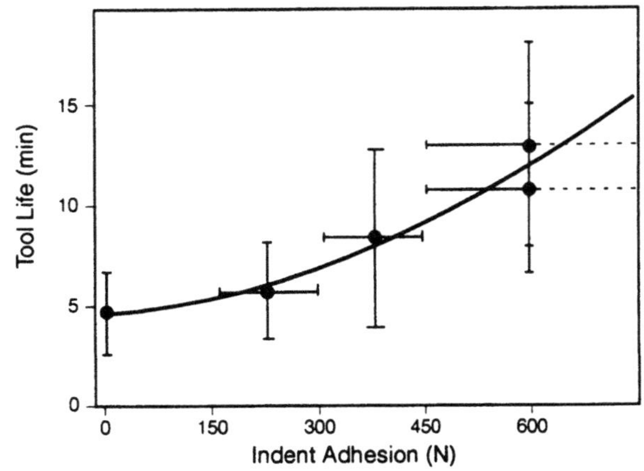

Fig. 12. Influence of indent adhesion on the life of PVD TiN coated carbide inserts.[26] Courtesy of the Mineral, Metal and Materials Society of AIME.

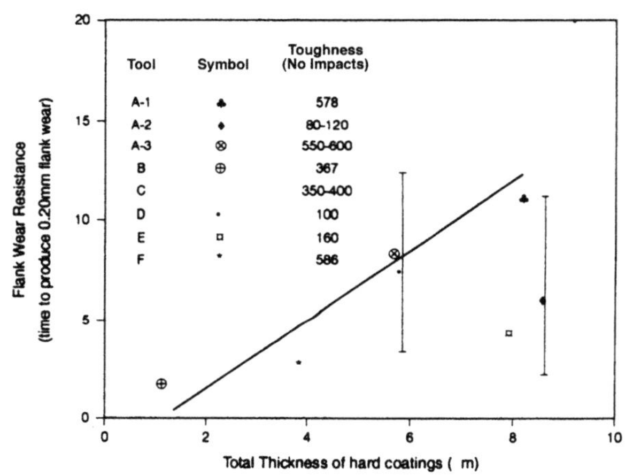

Fig. 13. Flank wear resistance of multi-layer CVD coated tool inserts.[27]

5. THE WAY FORWARD: TOWARDS THE ESTABLISHMENT OF NATIONAL AND INTERNATIONAL STANDARDS

From the foregoing it is evident that the establishment of agreed quality assurance procedures and standards in advanced surface engineering – advanced coatings – has not yet taken place for at least two main reasons: (i) the very rapid rate of change in the technology – PVD processing has effectively only been a commercial practice in the field of tribological coatings for not much more than a decade, CVD preceding it by about a further 10 years or so; and (ii) the relatively poor understanding which there is, especially at the fundamental level, concerning the properties and characteristics of advanced coatings, and their significance, especially with respect to design and performance of a coating system.

The ultimate objective would be the ability of the coatings user to select the best, optimised coating (system) for particular purposes, using some form of expert system derived on the basis of experience, and especially following selected in-depth research and development studies. The coater would act as an essential link between the user, commonly the manufacturer, including the die and tool maker, and the researcher, providing advice, on the basis of extensive experience, of the practicability of producing a particular coating system, taking into account the agreed total design parameters. As has been pointed out by Strafford *et al.*[21] there is an evolutionary pattern in the full systematic development of surface engineered components as set out in Fig. 14. Here the particular role of the experienced coater,

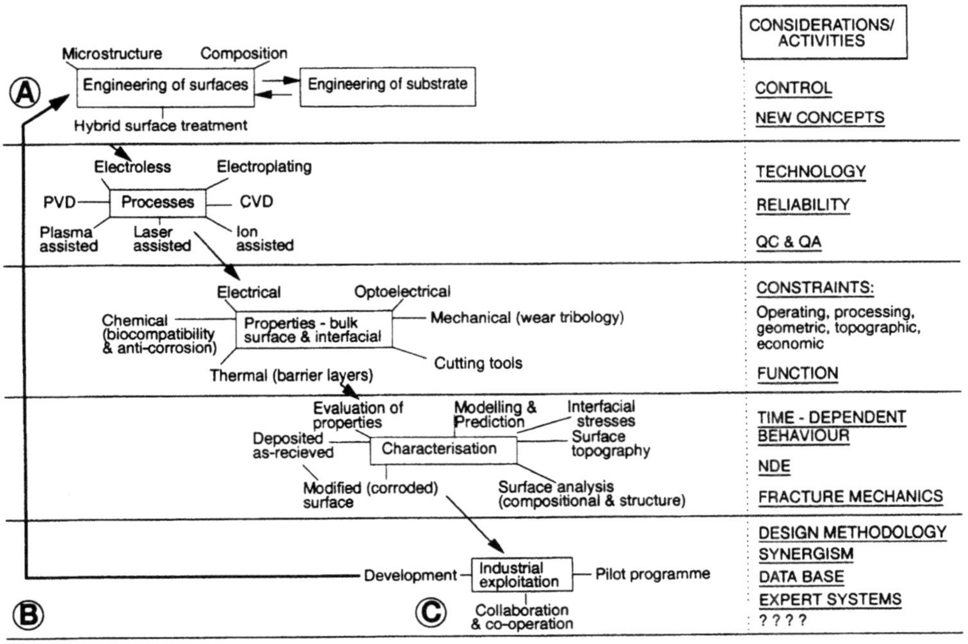

Fig. 14. Evolutionary development of surface engineered components.[2] Courtesy Ellis Horwood Publishers.

'at the coal face', and familiar with the problems of deposition of new hybrid coatings systems, including multilayer format, is seen as crucial. This may be particularly true in problems of scale up from laboratory to full-scale commercialisation.

There is also evidently a critical link between the researcher principally concerned with exploring, without prejudice, new coatings' systems involving new materials, and understanding their value and performance in the laboratory, and the coatings' user (the development engineer) confronting particular wear service conditions with engineering components. In particular, there is a need for coatings' performance in the laboratory to be correlated and understood with respect to the related (but no doubt differing in detail) behaviour in a full scale working situation.

In the move to develop quality and performance standards, then, there is a need to coordinate the activity of these various parties. In Australia there are a few scattered centres of excellence in R and D working in SE, serving key industries also in a few State centres. There is a need to establish a Working Committee on Advanced Coatings with appropriate representation (i) to ensure best practice is being followed with existing coatings; (ii) to advise on the newer emerging materials and systems; (iii) to monitor the take up and success with such new systems; (iv) to draw up, taking advice from appropriate quarters, appropriate quality standards and assurance procedures.

Such a Committee cannot and should not work in isolation. Extensive R and D and use of advances SE techniques is occurring in other parts of the world especially the USA and Europe. In Europe, in particular, the BRITE-EURAM pre-competitive research initiatives have funded much research including the surface technologies, drawing together academic establishments, R and D laboratories, and industry. Included in these programmes have been 'Round Robin' programmes of testing and evaluation, where experiences have been shared.

The recent agreement between the Australian and EC Governments, which allows Australian organisations now to join as full partners in Programmes, opens up new possibilities, including a common pursuit in the drawing up of agreed QA and QC guidelines for best practice in this crucial technology.

6. CONCLUSIONS

- In contrast to the more traditional areas of coating technology, where QC and QA standard are well established, the thin hard coatings such as TiC and TiN and other advanced multilayer coatings systems produced by CVD and PVD have not yet been the subject of systematic scrutiny and standardisation.
- While there has been much emphasis on the use of TiN and TiC for coated tools and dies to give enhanced wear characteristics, consideration of the background materials science suggests a potential in the use of many other compounds – oxides, nitrides, carbides and borides – which broadly have attractive intrinsic properties for use as tribological coatings.
- Some of the key aspects to successful advanced coating design using these materials may be identified, aspects which have been justified in performance in some circumstances, but much empiricism remains. There is a special need for a more systematic approach including mathematical modelling to address the possible use of alternative materials to TiN, with reference to specific wear situations.
- Although much effort has been expended in recent years in R and D programmes, especially concerning the preparation and performance of TiN coated dies and

tools, too little attention has been paid to understand fundamental structure/property relationships and performance indicators, in particular wear systems. This lack of knowledge, at the least, suggests optimisation may not have been achieved in the use of this particular compound. It has also hindered the establishment of appropriate QA and QC procedures, necessary for the satisfaction of the coating producer, and user, alike.

- There is a requirement to address the need for nationally and internationally accepted QA and QC procedures in the field of hard tribological coatings. The drawing up of such procedures will require active cooperation between the coating producer, the supply house and the manufacturer or user, and involve careful, structured and networked R and D programmes, with concurrent industrial trials. Without this systematic approach, the pace of this rapidly developing and critically important enabling technology may slow and its true potential may not be fully realised.

REFERENCES

1. *Coatings and Surface Treatments for Corrosion and Wear Resistance*, K. N. Strafford, P. K. Datta and C. Googan eds, Ellis Horwood, 1984, Chichester, UK.
2. *Surface Engineering Practice–Processes, Fundamentals and Applications in Corrosion and Wear*, K. N. Strafford, P. K. Datta and J. S. Gray, eds, Ellis Horwood, Chichester, UK, 1990.
3. K. N. Strafford, P. K. Datta and J. S. Gray: *Surface Engineering Practice–Processes, Fundamentals and Applications in Corrosion and Wear*, K. N. Strafford, P. K. Datta and J. S. Gray, eds, Ellis Horwood, Chichester, UK, 1990, Chapter 3.1.1, 397.
4. K. N. Strafford, C. Subramanian and T. P. Wilks: *Surface Engineering Vol. II Engineering Applications*, P. K. Datta and J. S. Gray, eds, Royal Society of Chemistry, London, 1993, **Paper 2.5.1**, 259.
5. K. N. Strafford, C. Subramanian and T. P. Wilks: *J. Mater. Proc. Technol.*, 1993, **38**, 431.
6. J. T. Burwell and C. D. Strang: *J. Appl. Phys.*, 1952, **23**, 18.
7. C. Subramanian and K. N. Strafford: *Materials and Design*, 1993, **14**, 1.
8. C. Subramanian and K. N. Strafford: *Wear*, 1993, **165**, 85.
9. H. Holleck: *J. Vac. Sci. Technol.*, 1986, **A4**, 2661.
10. U. Konig: *Surf. Coat. Technol.*, 1987, **33**, 91.
11. A. Matthews: *Surface Engineering Practice–Processes, Fundamentals and Applications in Corrosion and Wear*, K. N. Strafford, P. K. Datta and J. S. Gray, eds, Ellis Horwood, Chichester, UK, 1990, Chapter 1.1.1, 33.
12. H. Holleck and H. Schulz: *Surf. Coat. Technol.*, 1988, **36**, 707.
13. H. Holleck, Ch. Kuhl and H. Schulz: *J. Vac. Sci. Technol.*, 1985, **A3**, 2345.
14. H. Holleck: *Surface Eng.*, 1991, **7**, 137.
15. O. Knotek, D. W. Munz and T. Leyendecker: *J. Vac. Sci. Technol.*, 1987, **A5**, 2173.
16. U. Konig and H. Grewe: *Tibologie: Reibung, Verschleiss Schmierung*, Vol. 1, Springer, Berlin, 1981, 197.
17. T. Bell: *Metals and Materials*, 1991, **17**, 478.
18. P. J. Martin: *Materials Aust.*, June 1986, 11.
19. R. Smart: *Advanced Surface Analytical Techniques in Characterisation and QA of Coatings*,

Paper presented at Second International Conf. on Surface Engineering, Adelaide March 1994. In Conf. Proc.

20. C. S. Subramanian, K. N. Strafford, T. P. Wilks, L. P. Ward and W. McMillan: *Surf. Coat. Technol.*, 1993, **62**, 529.
21. H. Dagnall: *Exploring Surface Texture*, Rank Taylor Hobson, Leicester, 1980.
22. M. Codet: *Surf. Coat. Technol.*, 1991, **45**, 1.
23. D. S. Rickerby and S. J. Bull: *Surf. Coat. Technol.*, 1989, **39**, 315.
24. T. Sato, M. Tada, Y. C. Huang and H. Takei: *Thin Solid Films*, 1978, **54**, 61.
25. A. J. Perry, M. Grossl and B. Hammer: *Thin Solid Films*, 1985, **129**, 263.
26. D. T. Quinto and G. J. Wolfe *Protective Coatings – Processing and Characterisation*, R. M. Yazici, ed., TMS, 1990.
27. W. A. Bryant, F. B. Battaglia, B. K. Downey and K. E. Undercoffer: *Metal Cutting Performance of Multi-Layer Coated Tool Inserts*, 12th Int. Plansee Seminar 1989, Vol. 3, Reutte, Tirol, Austria, 187.
28. L. P. Ward, T. P. Wilks, C. Subramanian, K. N. Strafford and M. McPhee: *Characterisation and Mechanical Properties of TiN, CrN and TiAlN Coatings in Relation to the Performance of Tools used in Manufacturing Industry*, Paper presented at the Second Australian International Conf. on Surface Engineering, Adelaide March 1994. In Conf. Proc.

TOWARDS THE ESTABLISHMENT OF QUALITY CONTROL STANDARDS FOR ADVANCED COATINGS: AUSTRALIAN AND INTERNATIONAL PERSPECTIVES

G. O. MACINTOSH
Standards Australia, 19–25 Raglan St, South Melbourne, Victoria 3205, Australia

and

B. A. WILSON
Aeronautical and Maritime Research Laboratory – Melbourne

ABSTRACT

In anticipation of a future need for a national standard to cover testing and quality control aspects of high technology coatings including titanium nitride (TiN) coatings, the authors have set out this paper into the following two areas:

1. Aspects of quality control.
2. A brief history of the development of metallic surface coating processes. Included is a description of three types of conventional metallic coatings (electroless nickel, hard chromium and gold) each of which have some characteristics similar to those of titanium nitride coatings, e.g. wear resistance, corrosion resistance and resistance to galling. A table compares test requirements for these coatings as specified by a number of standards.

1. ASPECTS OF QUALITY CONTROL

If quality can be defined as fitness for use, then quality control is the broad task of maintaining a state of product and service attributes that assure a customer of a useful product and a manufacturer of a viable business enterprise.[1] This has particular meaning for metalworking, metal-finishing and metal-treatment companies. Too often, an existing specification, process description or control parameter is taken for granted without asking whether the process or specification will really result in customer satisfaction.

All products and services have variability because it is impossible for all units of a product or service to be made exactly alike. When a product is made with highly precise manufacturing equipment, variability may be difficult to measure, but it is always present; it is an effect to be observed and measured.

Part of the variability is beyond control; it is inherent in the process and materials. This portion of variability can be changed only by changing the basic process or material. The remainder of the variability is due to assignable causes – causes that can be reduced or eliminated if they can be identified.

In order to limit the variability of advanced coatings such as titanium nitride, advanced testing techniques which have evolved from conventional test methods are being developed. These techniques, with appropriate acceptance criteria, can become requirements of specifications or national standards.

2. A BRIEF HISTORY AND DESCRIPTION OF METALLIC SURFACE COATING PROCESSES

2.1 Electroplating

While some metal coating procedures date back to ancient times, modern electroplating started in 1800 with Volta's discovery of the voltaic pile, or battery which for the first time made significant quantities of direct current electricity available.[2] The battery was employed to deposit lead, copper and silver on themselves and on a variety of basis metals such as gold and iron.

Commercial electroplating of metal coatings began around 1840–41 mainly for decorative use and was accelerated by the discovery of cyanide solutions for plating silver, gold, copper, and brass. A cyanide–copper solution, for example, gave adherent deposits of copper directly on iron and steel. It is still used as a mask for heat treatment as well as the undercoat on zinc die castings. Decorative nickel plating was introduced in Boston in 1869 and achieved immediate popularity. The plating process required pure nickel anodes and salts in a neutral or slightly acid nickel solution.

Many innovations were introduced into electroplating practice around the turn of the century. Electrolytic cleaning and the bulk plating of small parts in tumbling barrels were introduced. Zinc plating of steel parts for rust protection gradually established itself, followed by cadmium plating for the same purpose around 1920. Tin plating was well known from the middle of the 19th century, but it was not until World War II that tin plating of steel strip for thin cans was carried out on a large scale. This process was more economical than the conventional hot dip method.

Other developments in the early 1930s and 1940s were focused on producing bright coatings which would reduce the amount of manual polishing required to produce a bright highly reflective finish. Plating solutions containing organic surfactants and brighteners to 'level' and 'brighten' the plated deposit were developed.

While the foregoing electroplating processes use aqueous solutions, some metals can be plated from non-aqueous solvents and from fused salts. Aluminium, for example, can be plated from either solvent-based electrolytes or from fused salts; the electrowinning of aluminium is the major example of plating from a fused salt process. Fused salt electroplating is one of the few ways of electrodepositing refractory metals, i.e. tungsten, molybdenum, tantalum and niobium, and while having very little practical application is a method of producing special wear-resistant coatings on metals.

Electroplated chromium became the major electrodeposited wear-resistant coating and revolutionised industry when introduced in 1925. It was a bright lustrous coating which was both decorative and corrosion resistant and had very good wear-resistant properties. While there were many attempts to discover a dependable chromium plating process, they were unsuccessful until 1924. Most of the early work had involved solutions of simple trivalent and bivalent salts of chromium to produce the chromium coatings. Even now in the 1980s and 1990s trivalent chromium plating baths are still the subject of intensive research to eliminate the use of environmentally unfavourable hexavalent chromium.

Chromium plate found a ready market in the automotive and appliance fields, in which the merits of the combination plate nickel–chromium or copper–nickel–chromium were soon proven. The requirements for closer control procedures for bath composition, temperature, and current density were reflected in better control and development of other processes.

The development of hard chromium plating created a new way of improving the wear resistance of machine parts and their operation due to its good frictional and heat resistance properties. Worn or undersized parts were built up with chromium plate.

The success of hard chromium plating is partly due to several intrinsic properties of the process and the coating. The highly oxidising solution of chromic acid (25%) plus sulphuric acid as a catalyst operating at 50°C is self cleaning on the base metal and produces very adherent deposits under almost all conditions; the process is easy to control and very tolerant of impurities. The chromium coating is deposited in a highly stressed state resulting in the formation of small cracks which are overcoated during subsequent deposition thus producing a total coating of very high hardness. The chromium deposits can cover machining defects in the surface of the basis metal. The decreased friction characteristics of chromium deposits is a result of intrinsic properties of the metal itself and the method of deposition.

While nonmetallic materials have been plated since the mid-19th century, a period of rapid growth in the utilisation of electroplated plastics began in 1963 with the introduction of ABS plastic (acrylonitrile–butadiene–styrene), which was readily plated.

2.2 Electroless nickel coatings (autocatalytic nickel–phosphorus coatings)

While autocatalytic deposition processes predate electroplating (the silvering of mirrors by the reduction of a silver salt was used in the early 1800s), the first autocatalytic nickel–phosphorus coatings were produced in 1844. Development of the nickel–phosphorus coatings was carried out at the National Bureau of Standards and the first trial commercial baths were introduced in 1955.

Electroless nickel–phosphorus coatings can be obtained from baths formulated to produce bright, semi-bright or dull deposits. Their appearance depends upon the brightness and smoothness of the substrate. The deposits are normally used without machining or polishing.

These coatings are produced by the catalytic reduction of nickel metal ions in hot (80–95°C), usually mildly acidic aqueous solutions at atmospheric pressure using sodium hypophosphite as the reducing agent. Suitable solutions, proprietary or otherwise, are those that produce a deposit which will meet the requirements of the product specification.

The coatings produced are uniform in thickness on irregularly shaped parts, provided that the processing solution has circulated freely over all surfaces.

The coating is a metastable alloy of nickel and phosphorus containing up to 15% (m/m) of phosphorus. The physical and chemical properties and the structure of autocatalytic nickel are dependent upon the coating composition, the chemical make-up of the plating bath, the pretreatment of the substrate, and heat treatment after plating.

Heat treatments up to 200°C used to improve adhesion or to give relief from hydrogen embrittlement do not impair the corrosion resistance, or substantially modify the hardness or the wear properties of the coatings.

An important property of electroless nickel coatings is the ability of the coatings to be hardened by heat treatment at temperatures up to 450°C. The hardness of the coating can be increased from 600 HV 'as deposited' to 1100 HV by heat treatment at 450°C.

2.3 Titanium Nitride Coatings

The commercial development of vacuum coating techniques started about 1950. Titanium nitride deposited by physical vapour deposition techniques is applied to improve surface properties, in particular wear resistance, surface hardness and to reduce the tendency for galling. The coating is resistant to corrosion under ambient conditions and to oxidation at temperatures below 450°C. It is particularly suitable for the protection of items which need to be handled and is compatible with many reactive materials.

Surfaces containing oxide, rust, black oxide scale, grinding burns or other machining marks, cracks or burrs are not suitable for coating. Processing conditions used for the coating of hardened and tempered steels must be carefully controlled to

Table 1. Test methods specified in standards covering electroless nickel, hard chrome, gold and titanium nitride coatings.

Test method	**Electroless nickel (ISO 4527)**	**Hard chrome (AS 2453)**	**Gold (AS 1901)**	**Titanium nitride (DEF STD 03–28 (Pt 3)**
Appearance				
roughness	X (ISO 468)		X	
regular Cr (Ferricyanide porosity (paper))		X		
crack-free Cr (Macro/micro method)		X		
porous Cr (Comparison diagrams)		X		
general				X
Thickness				
Average (strip and weigh)	X		X	
micrographic	X	X	X	X
coulometric	X	X	X	
magnetic	X	X	X	
magnetic induction and eddy current			X	
beta-backscatter	X		X	
direct measurement before and after plating	X			
SEM	X			
X-ray fluorescence			X	
Adhesion				
180° bend test	X	X		
thermal shock test	X	X	X	
punch test	X			
file test	X		X	
grind and saw test		X	X	
chisel test		X	X	
burnish test			X	
peel tests			X	
scribe and grid			X	
scratch test				X
Porosity				
Ferroxyl test	X			
microscopic test		X		
thioacetamide anti-tarnish and porosity test			X	
sulphur dioxide/hydrogen sulphide porosity tests			X	
electrographic porosity test			X	

Table 1. *Continued*

Test method	**Electroless nickel (ISO 4527)**	**Hard chrome (AS 2453)**	**Gold (AS 1901)**	**Titanium nitride (DEF STD 03–28 (Pt 3)**
Corrosion protection of substrates				
NSS test	optional tests			
ASS test				
CASS test				
TAA test (thioacetamide test)				
CORR test (corrodkote test)				
Corrosion resistance of coating				
NSS test	optional tests			
ASS test				
CASS test				
Wear resistance				
abrasive wheel	optional tests			
oscillating pin				
block and pin on ring				
solid particle impingement				
liquid impingement erosion				
vibratory cavitation erosion				
Solderability				
various methods	X			
Chemical composition				
various methods	X			X
Hardness				
Vickers and Knoop microhardness	X (ISO 4516)	X	X (ISO 4516)	
Electrical properties			X (relevant product spec.)	

prevent impairment of their metallurgical properties and structure. The presence of a thin titanium metal underlayer improves the adhesion and corrosion resistance of titanium nitride coatings. However, this underlayer may be omitted if its inclusion compromises the performance of the substrate/titanium nitride system as a whole.

Hardnesses in the vicinity of 3200 HV can be achieved on tools with a result that titanium nitride coatings on metal cutting tools can increase tool life by up to 10 times.

Quality control testing. Most tests required of titanium nitride coatings involve specialised equipment and include the following:

(a) Chemical composition: Methods include mass spectrometric, SEM–EDAX and X-ray diffraction.
(b) Thickness: Micro-sectioning.
(c) Adhesion: Precision scratch test.
(d) Hardness: Microhardness using Vickers scale.
(e) Visual examination.

A comparison between the quality control tests required by current standards on electroplated gold coatings (AS 1901–1993), electroless nickel coatings (ISO 4527–1987), hard chrome coatings (AS 2453–1981) and titanium nitride coatings (DEF STD 03–28(Pt 3)) is given in Table 1. Australian standards were selected, where available, in preference to overseas standards. The table is included to provide test method options for consideration should the development of a standard on titanium nitride coatings go ahead.

REFERENCES

1. R. A. Abbott: 'Quality Concepts and Statistical Methods', ASM Handbook Vol. 11, 8th edition, *Nondestructive Inspection and Quality Control*, 401.
2. Article on electroplating, *Encyclopaedia Britannica, Macropaedia*, Vol. 6, 1980.

STANDARDS REFERENCED

ISO 4527–87 Autocatalytic nickel–phosphorous coatings – Specification and test methods.

AS 1901–93 Electroplated coatings – Gold and gold alloys.

AS 2453–81 Electroplated coatings of chromium for engineering applications.

Def Std 03–28 (Part 3)/1 Physical vapour deposition of titanium nitride for surface protection.

QUALITY ASSURANCE OF ADVANCED COATINGS: THE COATER'S PERSPECTIVE

N. A. G. AHMED
Plasmion Ltd., Salford Tramways Building, P.O. Box 50, Frederick Road, Salford M6 6BY, UK

ABSTRACT
Quality management systems have become an essential part in the surface coating industry to improve the product and the handling procedures. When using the ion-assisted deposition technique to deposit different coatings on substrates, quality control procedures usually include the various process parameters to ensure high quality and reproducibility of the coating on an industrial scale.

1. INTRODUCTION

The increasing requirement for high-quality coatings has made it inevitable for the surface coating industry to put more effort into precisely controlling the coating process and other ancillary processes. Therefore, quality management systems have become an essential part in improving the product and handling procedures. International quality management standards within ISO9000 have already been adopted by 55 countries around the world.[1] Many companies, including aircraft and tool manufacturers, now require ISO9000 certification of outside vendors for most of their purchased and subcontracted parts, including surface treatment and coatings.[2] The quality system adopted by the surface coating centre ensure that customer's requirements are considered at all stages, guaranteeing good process control according to the particular specification, reliable performance, prompt delivery and efficient service. Accordingly, continuous monitoring and/or compliance with documented procedure is needed to ensure that the specific requirements are met consistently. This means that records have to be maintained for qualified processes, equipment and personnel, as appropriate. This usually assists in identifying the real reason for any failure as quickly as possible in order to take the necessary action to resolve the problem before shipment. Non-conforming parts can, therefore, be either recoated or returned to the customer. As part of the quality control schedule, procedure records are documented and signed by the relevant personnel. Inspection and test records should also be maintained throughout the whole process to ensure conformity to standards.[3]

The present paper reviews the quality control system currently utilised in the area of ion assisted deposition technique and describes the control of the various stages of the coating process indicating possible sources of problems.

2. PROCESS CONTROL

To attain high quality reproducibly, many parameters have to be adjusted and controlled for each Ion Assisted Deposition system. Although it is not possible to control all process parameters independently, most of them can be monitored and controlled to certain limits depending on the flexibility of the system employed. Reactive systems, in particular, require a high degree of independently-controlled parameters to achieve the required reaction and the final surface finish.[4] Therefore, the presence of an active gaseous discharge introduces additional constraints as some of the variables controlling the process parameters also affect the plasma variables. In general, there are three major processes that have to be monitored in surface coating technology to obtain high quality coatings reproducibly. These are pre-treatment, the ion-assisted deposition process and post-treatment whenever required.[5] Each

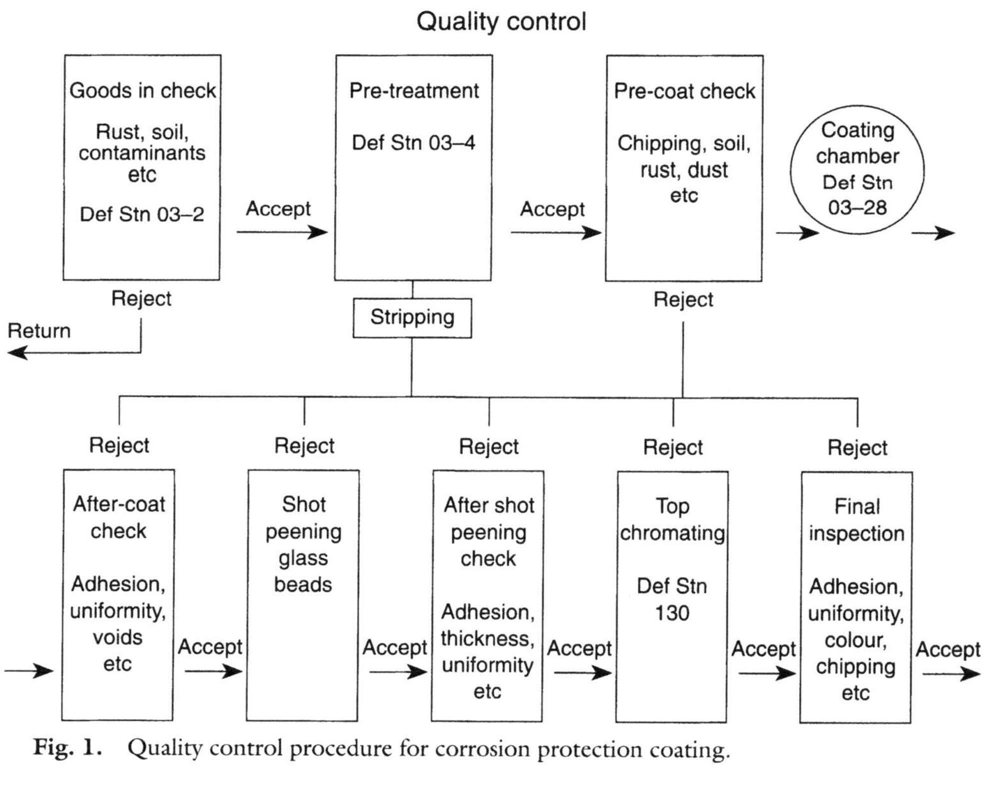

Fig. 1. Quality control procedure for corrosion protection coating.

process has its own set of parameters that have to be adjusted independently. Figure 1 sets out the various stages of a quality control procedure to conform to the specifications for corrosion protection coating on aircraft parts. Figure 2 shows some

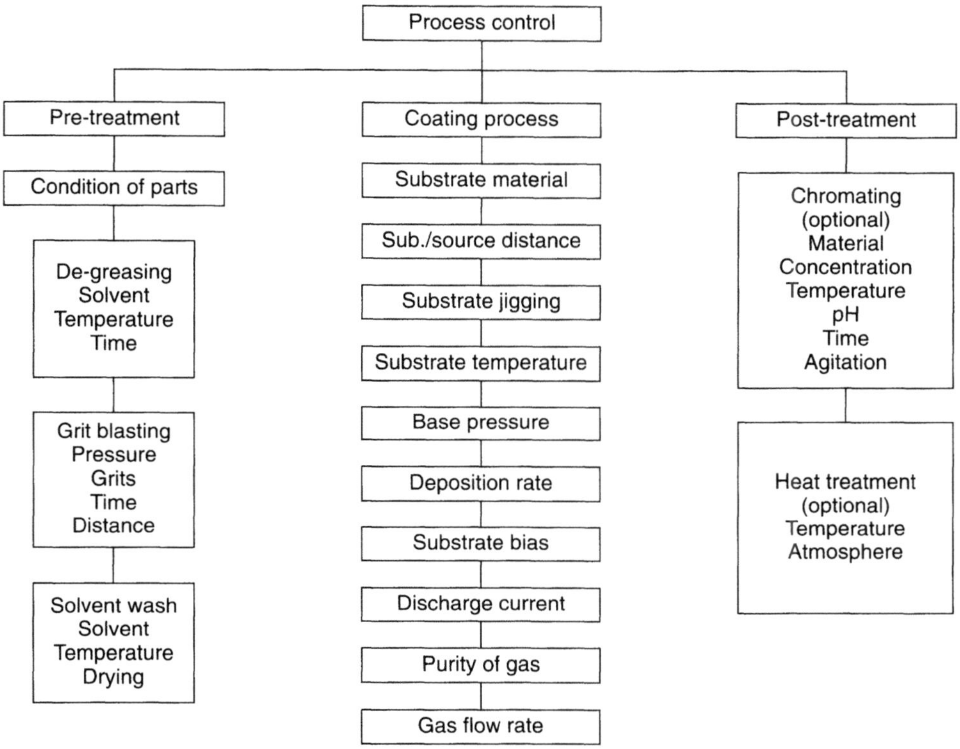

Fig. 2. Process parameters that can influence the quality of IAD coating.

of the most important parameters that can influence the quality of the final coating. Understanding the inter-relation between the various parameters can lead to a better optimisation of the plasma deposition process. While some of these parameters can be monitored during the plasma coating process others can be monitored at each stage of pre- and post-treatments.

3. PRE-TREATMENT PROCESS

The first important stage to implement the quality control system is the pre-treatment of the parts prior to coating. During this stage parts received are thoroughly inspected for any sign of rust, contamination and damage before cleaning. Cleaning is generally defined as the process of removing undesirable matter from the surfaces of the parts prior to coating. It usually involves the breaking of

ION VAC TECHNOLOGY LIMITED
UNIT 6, WILLAN INDUSTRIAL ESTATE, VERE STREET,
SALFORD, M5 2GR
Tel: 061–745 9598

STANDARD PRE-CLEANING RECORD

DATE	TIME	BATCH	OPERATOR	PART TYPE	QUANTITY

PART	TRICH TIME	GRIT TYPE	GRIT PRESSURE	IPA TIME	APPEARANCE*

Batch passed to coating at..........................hrs on .... by ....

IVT 4

NOTE:-
*APPEARANCE - List whether bright or dull and if there is any chipping

INSTRUCTIONS:-
1) Insert in wallet and issue with batch
2) When batch is completed form remains in wallet and is filed

Fig. 3. Quality control pre-cleaning record.

bonds between contaminant particles and the substrate surface by utilising mechanical or chemical methods.[6] The selection of a suitable cleaning process depends mainly upon the nature of the substrate being cleaned, the soil to be removed and the size and number of parts to be cleaned. While light polishing, scrubbing and ultrasonic cleaning with detergents or solvents may be adequate in the laboratory, harsh treatments with grit blasting and chemical etching will be required in industrial production.[7] Therefore, the removal of such soils requires more than one cleaning operation to achieve the degree of cleanliness necessary for producing good coating quality. Figure 3 shows an example of the quality control schedule that has to be filled in by the operator during the stage of pre-treatment. Figure 4 shows

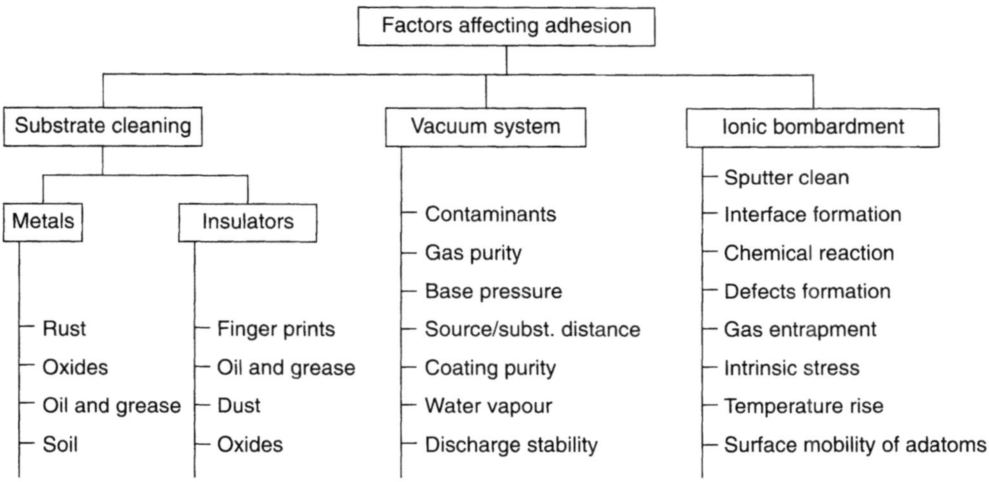

Fig. 4. Factors that can influence the IAD coating adhesion.

factors that can influence the final adhesion of the coating due to substrate cleaning. Inadequate surface cleaning and non-uniform grit blasting usually yield areas with bad adhesion on the coated parts.[8]

4. QUALITY CONTROL DURING IAD PROCESS

The cleaned parts should be moved to the vacuum chamber within 30 minutes otherwise the parts may require re-cleaning. In the coating cycle the control of various parameters such as vacuum level, rate of outgassing, uniformity of plasma bombardment, etc. (see Fig. 5) ensures the quality of the coating with minimum adhesion failure. At this stage, cleaning is carried out by direct ion bombardment which leads to substrate heating. It also involves other mechanisms, such as disorption of impurities from the surface by electrons, neutrals or ion bombardment, or volatisation of organic residues by chemical reaction with dissociated oxygen.[9] When oxygen is used for sputter cleaning, care must be taken not to oxidise the

ION VAC TECHNOLOGY LIMITED
UNIT 6, WILLAN INDUSTRIAL ESTATE, VERE STREET,
SALFORD, M5 2GR
Tel: 061–745 9598

STANDARD FIRST COAT VACUUM COATING RECORD

DATE	TIME	BATCH	OPERATOR	PART TYPE	QUANTITY

A) PARAMETERS:-

ULTIMATE PERIOD	CLEAN PRESSURE	CLEAN AMPS	CLEAN VOLTS	CLEAN TIME

COAT PRESSURE	COAT AMPS	COAT VOLTS	COAT TIME	PASSES	COOL TIME

B) LOADING REQUIREMENTS (support plates viewed from above):-

A) SADDLE Top Surface	B) ADAPTOR Holed End	C) SCREW Threaded End
C* B* C* D* E* F*	A* B* C* D* E* F* G* H* J*	A* B* C* D* E* F* G* H* J* K* L* M* N* O* P* Q* R* S*
Use Small Collet (773);	Use Adaptor (776);	Use Rod (778).
Washer (774A); Rod (772)	Rod (777A).	
Use M5 fixing holes		

C) PART NUMBER LOCATION

A ....; B ....; C ....; D ....; E ....;
F ....; G ....; H ....; J ....; K ....;
L ....; R ....; S ....;
Q ....;

First coat completed at..............................hrs on .... by ....

IVT 5-1

INSTRUCTIONS:-
1) Insert in wallet and issue with batch
2) When batch is completed form remains in wallet and is filed

Fig. 5. Quality control IAD process record.

substrate and the evaporated or sputter material. For cleaning insulated components, such as glass or plastic, the parts can be biased with an rf power or can be separated from the flow discharge circuit. The control of evaporation or sputtering rate particularly during reactive coating is very important to achieve the required chemical stoichiometry. In reactive coatings, the chemical reaction may take place at one or more of three possible sites in the coating chamber. The reaction may take place in the vapour phase, the substrate, coating surface, on the evaporant or target surface. The bulk of evidence, however, suggests that reaction occurring at the substrate surface is the most important mechanism in the formation of a compound coating.[10] Therefore, the control of the gas flow mixture, the evaporation or sputtering rate, substrate temperature and the ionic bombardment are the most important factors in reactive deposition. The location of the gas inlet around the substrate is also an important parameter in reactive coating. Figure 5 shows an example of the process record sheet which logs in all information regarding the coating cycle and the location of the coated parts inside the vacuum chamber. This assists in tracing back any changes in the uniformity of the coating and problems occurring during the coating process.

5. POST TREATMENT

Certain types of coatings may require some form of post treatment prior to being used in actual service. Heating the parts to relieve stresses or top chromating are the most common forms of post treatment.[11] Heat treatment may take place inside vacuum ovens, whereas chromating requires chromium solution to add an additional protective layer on the top of aluminium, cadmium or zinc coating for extra corrosion protection. The supplemental chromate treatment is also applied for the adhesion of subsequent painting applications. Such treatments usually require some degree of quality control to ensure the control of temperature and time in the case of heat treatment and the solution variables such as temperature, pH, concentration and immersion time in the case of chromating.

6. TREATMENT OF REJECTS

Parts that are not conforming to the specific standards are either stripped and recoated or returned to the customer for further machining. Treatment of the rejects rely on the mutual discussion and agreement between the coating firm and the customer to avoid any unexpected problems particularly relating to tolerance and substrate finish. In most cases, however, rejected parts can be stripped using chemicals or grit blasting then recoated using the same procedure as for new parts. The stripping and recoating procedures should be logged in the process sheet as part of the quality control system. The use of chemicals such as acids for stripping the rejects may produce hydrogen embrittlement which consequently affects the corrosion protection of the coating during service.[8] Grit blasting at specified air pressure can be used for stripping the rejects but care must be taken not to exceed the limit of tolerance particularly on threaded parts. In some cases, the coating may react with the substrate material to produce brittle compound that can be easily removed during grit blasting. As an example, stripping rejected cadmium coated threaded parts with grit blasting at 50 psi air pressure may remove 60 micron. Only a few microns of uncoated substrate material is removed following grit blasting at the same air pressure.

7. HEALTH AND SAFETY

Health and safety has become a vital issue in the surface coating industry due to the increasing use of various materials and compounds, some of them very toxic. Local health and safety authorities have special guidelines to follow for the use and disposal of different toxic materials and solvents. However, the management of the surface coating firm has also an important role in implementing all health and safety regulations on daily basis. If toxic materials are used then recommended respirators and overalls have to be worn to protect the employees from inhaling toxic dust or fumes. Good housekeeping on a daily basis should also be observed to conform to the health and safety procedures that limit exposure to toxic dust and fumes.[12] Measures must be taken when using chemicals and grit blasting materials to conform to Health and Safety Standards. It is also recommended for safety and quality that vacuum coating equipment is housed in a class 1000 clean room or other dust-free environment under an extraction hood to minimise the level of dust and fumes that can affect the quality of coatings and the health of operators.

8. CONCLUSIONS

The implementation of the quality control procedure in ion-assisted deposition technology can be easily measured by product yields, quality of product and minimum reject rate. This, in turn, will be reflected in customer satisfaction due to improved product quality and minimum quality testing from the customer side (Fig. 6). In addition, ensuring high quality results is a significant boost of confidence and productivity which means greater profitability. Management, however, has an important role in implementing quality control including health and safety procedures by providing staff with proper training on recognising all problems relating to the final finish and in monitoring the process control parameters with minimum supervision. Therefore, major problems related to adhesion failure, bad finish and poor performance can be identified and eliminated at early stages without affecting productivity and quality.

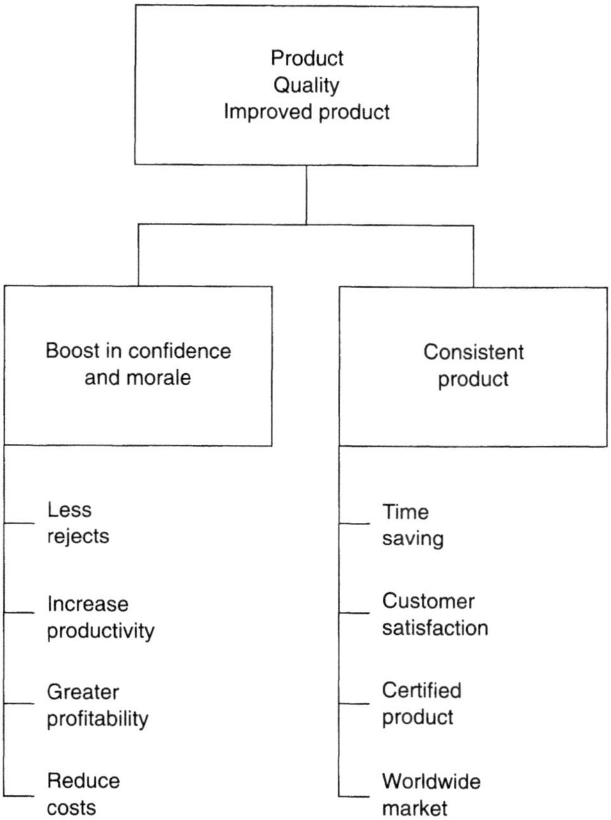

Fig. 6. Advantages of quality control system in Ion Assisted Deposition technology.

REFERENCES

1. R. McDonnell: *European Seminconductor*, June 1993, 21.
2. H. E. Gautheir: *Phototonics Spectra*, January 1993, 77.
3. BS Quality Systems, Pt 2 *Specification for Production and Installation* BS 5751 Pt 2. ISO9000 (1987). Published by British Standards Institution, London.
4. R. H. Bunshah and C. V. Deshpandey: *J. Vac. Sci. Technol.*, 1985, **A3**(3), 553.
5. N. A. G. Ahmed: *Product Finishing*, June 1991, 23.
6. D. M. Mattox: *Adhesion of Thin Films, Thick Films and Bulk Coatings*, ASTM STP 640, K. L. Mittal ed., American Society for Testing and Materials, 1987, 54.
7. N. A. G. Ahmed: *Ion Plating Technology, Developments and Applications*, Wiley & Sons, 1987.
8. *Tool and Manufacturing Engineering Handbook*, C. Wick and R. F. Velleux, eds, 4th Edn, Published by the Society of Manufacturing Engineering, Michigan, USA, 1985, Ch 16, 16–83.
9. H. K. Pulker: *Coatings on Glass*, Elsevier Scientific Publishers, Amsterdam, 1987.
10. *Thin Film Processes*, J. L. Vossen and W. Kern, eds, Academic Press, NY, 1978.

11. MIL–C–83488 (USAF) *Coating Aluminium, Ion Vapour Deposition*, 1976 and British MoD DEF STAN 03–28 Pt 2, *Cadmium Vacuum Coating*, 1993.
12. *Health and Safety Executive Guidance on Cadmium*, HMSO AB, ISBN 0 118839306, 1986.

VARIABILITY IN COATINGS AND TEST RESULTS

A. MATTHEWS and A. LEYLAND
Research Centre in Surface Engineering, University of Hull, Hull, HU6 7RX, UK

ABSTRACT
The problems of establishing standardised tests for coatings are discussed. In particular the variability in results from nominally identical coatings are highlighted, using as a basis a round robin trial carried out at three European laboratories. The properties assessed were thickness, hardness, adhesion and friction/wear performance. The test methods were the ball crater, Vickers microhardness, the scratch test and the pin-on-disc. The implications of variability are discussed and the usefulness of the methods is highlighted.

1. INTRODUCTION

A key factor in the quest to encourage the wider use of coatings by design engineers and others who make materials choices is to have a specification by which they can be denoted. This in turn requires that a standardised set of tests is available to confirm conformance with type. Particularly for the latest hard coatings, the achievement of such standard specifications has not been easy to achieve. This is partly because of the many deposition techniques available, which impart slightly different characteristics in each case; this prevents a 'process' based standard being applied. This means that the properties of the coatings must be the basis of the specification, and that in turn has led to severe problems in finalising agreed tests and values.

Many people make the mistake of assuming that surface properties such as hardness and wear rate should have exact and unique values for a particular coating. In practice such properties are influenced by surface defects and other variables which make them subject to statistical spread, like many other failure-related phenomena. Hardness is a particularly misunderstood property. As conventionally measured it is defined by the residual impression left in a surface when an indentor has been loaded into it and then removed. It would be to totally exaggerate the precision of the method or the simplicity of the deformation behaviour to assume that identical readings would be obtained from repeated measurements taken in the same vicinity on a surface. This is even without considering the subjectivity which goes into the assessment of indentation dimensions. Similarly with wear; this is a complex failure mechanism involving the interaction of many possible effects. To assume that these will be exactly repeated in separate tests is to misunderstand the nature of the processes involved – just in the same way that no knowledgeable person would expect nominally identical light bulbs to fail after exactly the same period of time. We must realise that in assessing coating properties we are often dealing with statistically varying effects and any attempt to standardise test methods must take this into account.

By way of illustration of the above points we shall refer to round robin tests carried out as part of an EC BRITE/EURAM project, first reported in Ronkainen *et al.*[1] Also we shall refer to more recent tests, especially those in Ollivier and Matthews.[2]

2. ROUND ROBIN TRIALS

2.1 Coatings Studies

As part of a project to develop new PVD ceramic coatings, three laboratories produced and evaluated films. The three laboratories were:
University of Hull (UH).
Technische Hochschule Darmstadt (THD).
Technical Research Centre of Finland (VTT).

For the purposes of this work, the laboratories involved each produced a different coating type, using advanced plasma-based deposition methods. The coatings were designed to be developments of the, now common, first generation coating, titanium nitride; the ones selected were ternary TiBN and TiAlN compounds. The combinations can be summarised as follows:

Source material	**Partner (coating designation)**	**Deposition method**
TiB_2 + BN	UH(A)	DC ionisation assisted EB PVD
TiB_2 + N	THD(B)	RF magnetron sputtering
TiAlN	VTT(C)	DC ionisation assisted EB PVD

Several coating thicknesses were included in the range 2–8 μm, representing typical values currently used for PVD deposits. The substrate material in each case was polished ASP 23 tool steel, hardened to 64 R_C, with a surface roughness of 0.4 μm Ra.

2.2 Test Methods

2.2.1 Thickness

Coating thickness measurements were carried out using the ball crater method. This involves the production of a spherical impression in the surface, as shown in Fig. 1. This was achieved using a steel ball, rotated on the surface, with a suspension of diamond particles in kerosene as the abrasive medium. The thickness of the coating is to a close approximation, given by the simple formula:

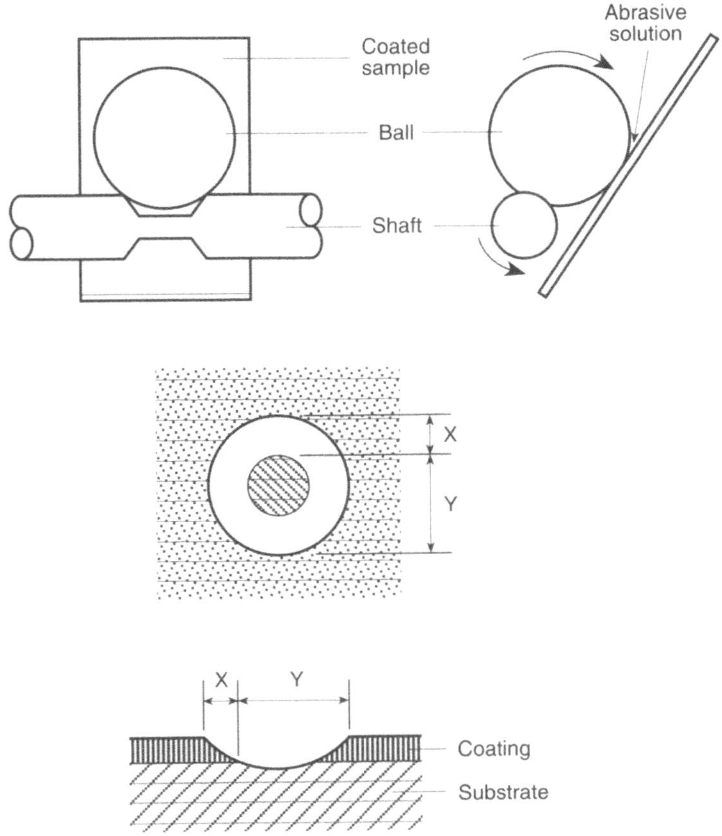

Fig. 1. Schematic views of the ball crater apparatus and the crater profile.

$$\text{Thickness} = \frac{x\ y}{\text{Ball diameter}}$$

UH and THD used a 25.4 mm diameter ball, whilst VTT used a ball 30 mm in diameter. Measurement of x and y were made at UH using a Shinko profile projector, at VTT using an Olympus BH-2 microscope and at THD using the measurement facility on a Leitz Miniload microhardness tester. Each laboratory produced one crater from which four determinations of thickness were made to obtain mean and standard deviation values. All craters were located within a 4 mm × 5 mm area on each sample. Further details of the ball crater technique are given in Valli *et al.*[3]

2.2.2 Hardness

For thin coatings, of the type evaluated here, a micro-indentation hardness test is preferred. This ensures that the influence of the substrate on the readings is minimised. Often a Knoop indenter, which produces an elongated impression and has a shallower penetration, is used, but in these tests we used the more common Vickers microindenter. VTT used a PMT 3 machine, UH and THD used Leitz Miniload microhardness testers. In all cases loads of 15, 25, 50 and 100 g were used. The indentation dwell time was 10s. The mean and standard deviation of the hardness number was determined from each group of five indentations by one person in each laboratory.

2.2.3 Adhesion

A scratch test was used for assessment of the relative adhesion of the various coatings (see schematic diagram in Fig. 2). VTT and UH used scratch testers manufactured

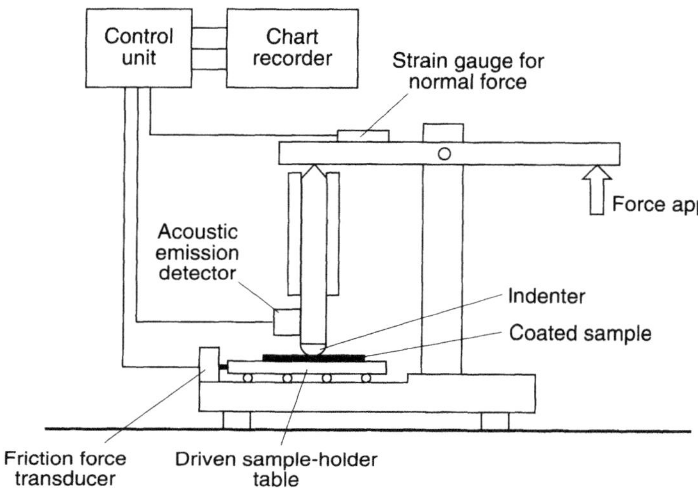

Fig. 2. Schematic view of a scratch tester.

by VTT Tech. THD used a CSEM (LSRH) Revetest machine. All three machines were equipped with acoustic emission and friction force measurement. The loading rate was 100 N min^{-1} and the table speed was 10 mm min^{-1}. The nominal maximum load was 100 N. Each machine used a Rockwell 'C' diamond indenter geometry (radius 200 μm). The indenter was cleared of residues by manually polishing against a hard titanium nitride coating (on a high speed steel substrate), and subsequently examined in an optical microscope to confirm removal of debris. As for hardness testing, the tests were carried out under ambient laboratory temperature and humidity conditions (typically 20°C, 50–70% relative humidity). Three scratch tests per coated sample coupon (one near each end and one near the centre) were performed at each laboratory. The mean and standard deviation of 'designated' critical loads was evaluated from each group of three scratches. A typical friction force output is shown in Fig. 3.

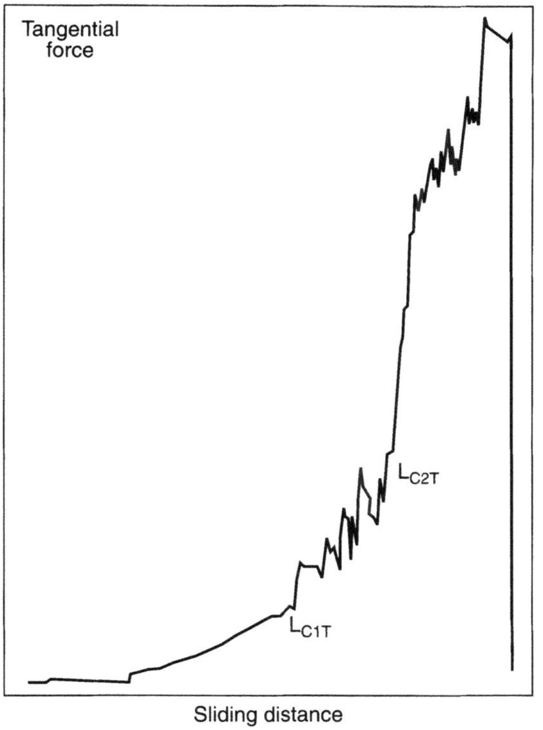

Fig. 3. A typical tangential force trace from a scratch test.

2.2.4 Friction and Wear Performance

There are a great many machines available to assess tribological performance of coatings. The one selected for this work was a 'pin-on-disc' configuration, which is expected to be proposed as a European Standard under the VAMAS scheme. The tests were carried out using a fixed 10 mm diameter polished ball as the 'pin'. The 'disc' was mounted with its diameter in the horizontal plane on the VTT and UH machines, and in the vertical plane on the THD machine. The pin material was M50 steel (80 Mo Cr V42 16), nominal hardness $62R_C$. The discs were coated, being ASP 23 material (nominal hardness $64R_C$). The tests were performed unlubricated at a sliding speed of $0.1\,m\,s^{-1}$, and a normal force of 10 N; they were run for a sliding distance of 250 m. The relative humidity during testing was controlled to 50 ± 2% at VTT and UH; at THD the ambient humidity was 68 ± 4%. The temperature for the VTT and UH tests was 21 ± 2°C, whilst for THD it was 23 ± 1°C.[1]

2.3 Results

2.3.1 Thickness

Figure 4 shows the mean thickness values for each coating found from ball cratering, the error bars represent standard deviations. Agreement between the mean thickness

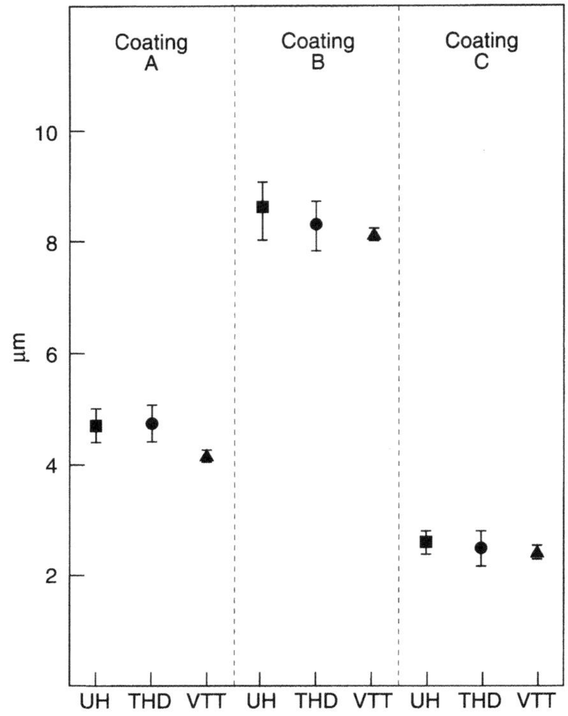

Fig. 4. Coating thickness measurements.

values was good. In all cases, the VTT measurements have the lowest standard deviation values, which is attributed to the ball craters being more symmetrical than those from UH and THD. The coating thickness or surface structure did not have any effect on the consistency of results.

2.3.2 Hardness

Figures 5(a), (b) and (c) summarise the Vickers microindentation hardness results as a function of load for each coating. The error bars represent standard deviations. Probably the most interesting aspect is the considerable difference in absolute values from each laboratory. This can be most clearly seen in Fig. 5(c), where standard deviations for individual results are relatively small. Results from all three laboratories show a comparable hardness–load trend for coating C (Fig. 5(c)), which was the thinnest, but this is less apparent for coatings A and B (Fig. 5(a) and (b)). In the case of coating A, indentations were more difficult to discern, due to the surface structure of the coating.

2.3.3 Adhesion

In scratch testing the normal force was increased continuously. Acoustic detection and monitoring of tangential force, as well as optical detection, were utilised in determination of 'failure' events. The interpretation of the results was found to be

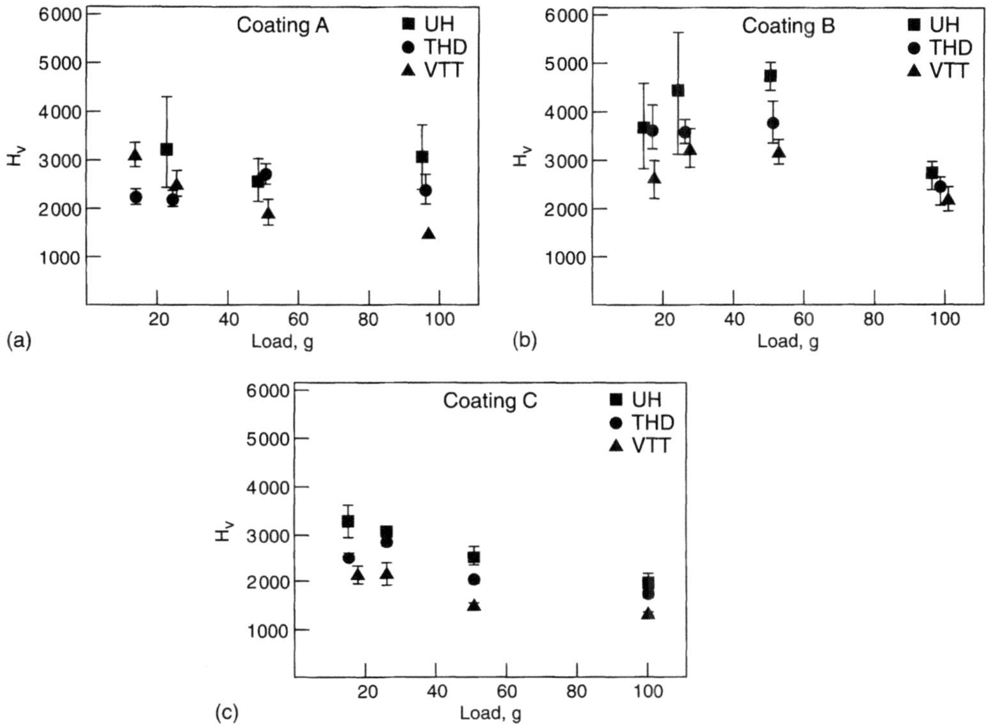

Fig. 5. Coating hardness results.

somewhat subjective. In particular, the definition of critical loads initially differed from laboratory to laboratory. After discussion between the laboratories, criteria for two critical loads L_{C1T} and L_{C2T} were determined from the scratch test tangential force traces. As shown in Fig. 3, L_{C1T} corresponds to the first significant change in the tangential force, and L_{C2T} to the first substantial change in the tangential force gradient. The results utilising these criteria (Fig. 6(a) and (b)) show a general agreement between laboratories within the standard deviation limits for both critical loads. No systematic correlation was found between results derived from acoustic emission traces from any of the laboratories. In some cases an acoustic emission signal was detected right from the beginning of the test, even though no other signs of the 'failure' events were detected. This was the case particularly with coating B.

Direct inspection of the scratches produced allowed the definition of two further critical loads, L_{C1D} and L_{C2D}. L_{C1D} is the first point of cohesive failure (failure within the coating) and L_{C2D} is the first point of adhesive failure (failure at the coating/substrate interface). L_{C2D} occurred within the scratch channel for coatings A and B, whereas coating C first showed adhesive failure outside the channel. Figure 7 summarises the L_{C1D} and L_{C2D} values. There is general agreement between the results of all the three laboratories within standard deviation limits. However, the L_{C1D} values of coating A show larger variations. The nature of the failure was similar,

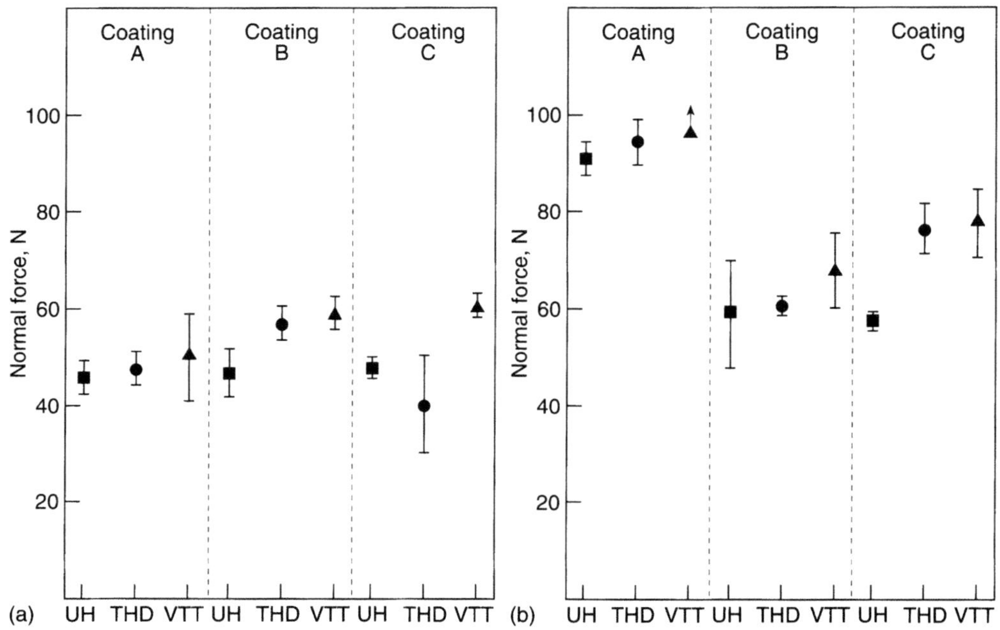

Fig. 6. Critical loads assessed from the frictional force: (a) lower critical loads, (b) upper critical loads.

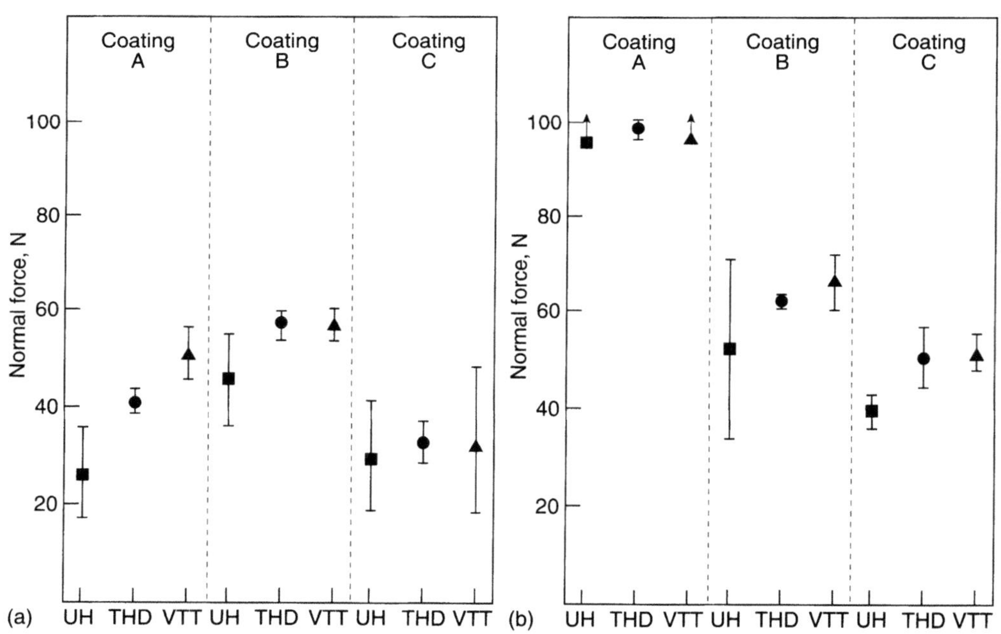

Fig. 7. Critical loads assessed by direct observation: (a) lower critical loads, (b) upper critical loads.

although it occurred sooner on the UH machine for this coating. When comparing the results derived from the friction output with those derived from direct observation, good agreement for coatings A and B is observed. However, for coating C the results are not comparable, the failure mechanism being very different.

2.3.4 Friction and Wear

The pin-on-disc tests were carried out twice in each laboratory for each coating. Figure 8 shows the values of the steady state coefficient of friction for the coatings in

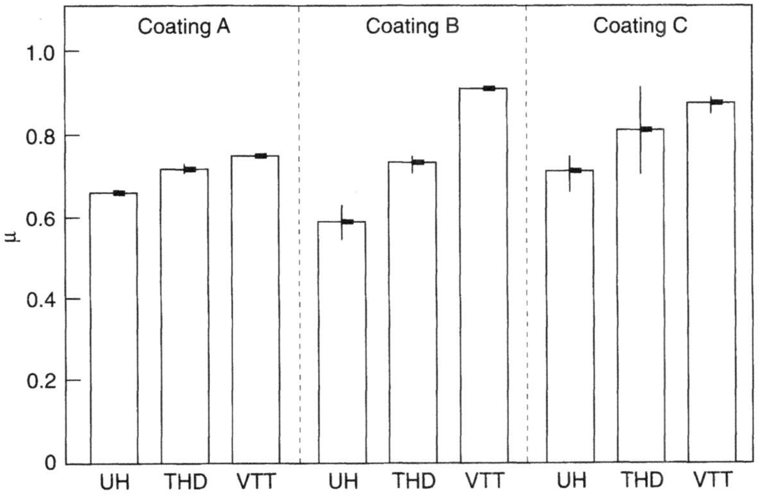

Fig. 8. Steady state coefficients of friction in pin-on-disc tests. Error bars represent standard deviations.

pin-on-disc tests. For coatings A and C the correlation is good, whereas for coating B the variation in coefficient of friction is from 0.59 to 0.91. Results show that the lowest values were measured in UH and the highest in VTT. This may be an indication of differences in the characteristics of the measurement equipment. However, the variation of the results is similar to that reported in other multi-laboratory studies carried out with bulk materials.[4,5]

The mean wear rate values and standard deviations are summarised in Fig. 9. The wear rates of the coated discs follow the same order as the steady state coefficient of friction values. Even though the standard deviations are quite large, the wear rates of the tests carried out in different laboratories are of the same order of magnitude. The wear rates for steel pins and coated discs for coatings A and B are in agreement, whereas the tests carried out in THD on coating C give higher values compared to the others. This is probably due to the influence of relative humidity at THD rather than to the mounting arrangement (i.e. the vertical disc).

During the tests, the formation of a white powder-like wear debris was observed

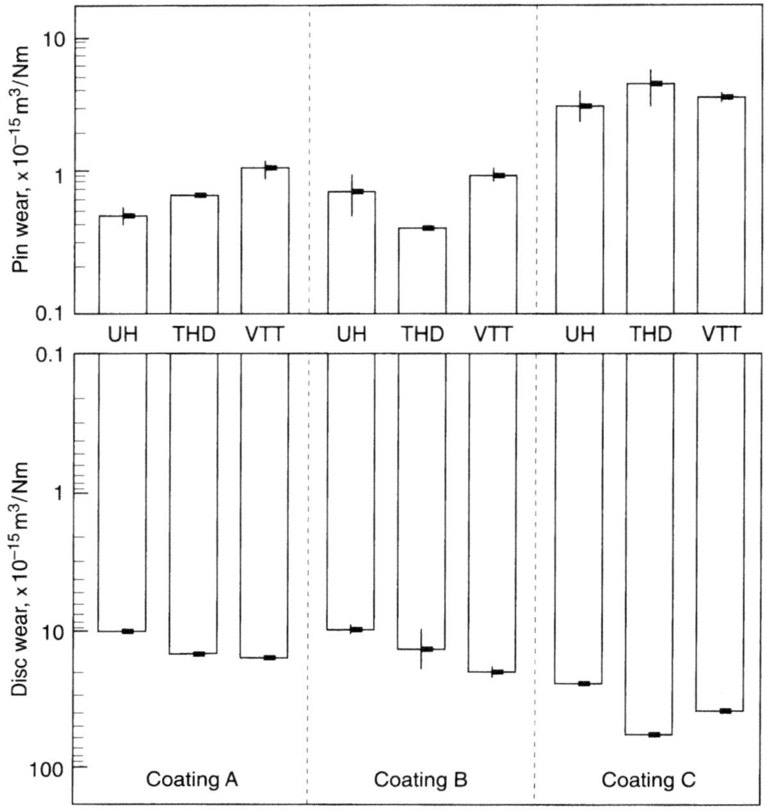

Fig. 9. Mean values of wear rates in pin-on-disc tests. Error bars represent standard deviations.

for all the coatings. Microscopic examination of the wear surfaces showed formation of transfer layers of reaction products on the wear surfaces of the pin in all cases. The wear surfaces of the discs were generally smooth with some evidence of scratches. The dominant wear mechanism seemed to be the same in each of the tests. In order to have nominally similar sliding speeds and distances, it was necessary to use different wear track diameters. This meant that the number of wear passes differed; UH being the greatest, followed by THD and then VTT. There was certainly no systematic increase in wear rate when comparing laboratories using increasing numbers of passes. This seems to confirm the importance of tribochemical processes compared to ones based on fatigue alone.

2.4 Further Comments

The work described above provided a useful insight into tests which were previously assumed to be relatively reliable and consistent. Indeed the ball crater for coating thickness measurements, micro-Vickers for hardness, scratch testing for adhesion and pin-on-disc testing for wear and friction characterisation, have been shown to provide useful information on the mechanical and tribological properties of the coatings. The use of these methods can be the basis of standardised coating evaluation methods but should always be treated with caution.

The ball crater test proved to be quick to carry out and its accuracy was good. The results are independent of factors such as hardness, thickness and surface roughness, in the range studied here. The two main real disadvantages of the test are removal of the coating on a small area of the surface, and the need to (ideally) have a flat region in which to produce this impression.

In Vickers microhardness testing the influence of the coating thickness is critical. The thinner coatings, like coating C, showed typical load–hardness dependence, where higher loads give a disproportionately lower measured hardness due to the plastic deformation contribution from the softer substrate material. The structural effects of the coating can make the indentation sizes difficult to measure, as in coating A, which can cause large variations in the results. It should also be noticed that for hard coatings, small changes in indentation sizes cause large differences in hardness readings. Therefore it is essential to include standard deviations with hardness values and always take account of the load employed.

In the pin-on-disc tests the variations of the results were considerable even though the same wear mechanism was dominant in all tests. In some tests the differences in relative humidity may have had some effect on the coating behaviour. Therefore there is a need to ensure identical test conditions. In practical situations complex interactions can take place between the failure mechanisms which cause performance variations. This should be taken into account when assessing new surface engineering developments and designing and optimising tribological contacts.

In the case of the scratch adhesion test, the L_{C1} (lower value) and L_{C2} (higher value) are suitable parameters to be used for cohesion and adhesion evaluation. Depending on the coating and substrate properties, the L_C values determined from the friction traces would be suitable for practical quality assurance of the coatings, whereas direct examination is useful for analysing the physical phenomena produced by the contact. In using the scratch test it is however important to recognise that many authors have commented that the results often do not correlate with the practical in-service performance of the coating. This is because the nature of the contact, where a hard diamond indenter plastically deforms the coating and substrate, is very different from the bulk elastic contact existing between (for example) two coated steel contacts. It has been suggested[6] that multi-pass scratch testing at low loads would better simulate real operating contacts, and there is some evidence that this is the case.

Also it has been pointed out[2] that, whilst the linearly increasing load scratch test samples a point at any particular load, a more ‘practical’ test such as the pin-on-disc will sample a length, or many points. This has led to the recommendation that when interpreting scratch test data, it is best to do many, and take the lowest (rather than the mean) as the value which will relate to failure in a practical situation.

3. DISCUSSION AND CONCLUSIONS

In an unpublished survey, carried out in the U.K. for the National Physical Laboratory, A Matthews and P. Holiday of Hull University identified the ten properties listed below as the most important to people using coatings on products.

1. Adhesion.
2. Thickness.
3. Surface finish.
4. Corrosion resistance.
5. Friction and wear performance.
6. Hardness.
7. Porosity.
8. Composition.
9. Internal stress.
10. Structure/morphology.

These represent the results from several dozen questionnaire responses and reflect a wide range of industry sectors. Whilst techniques for surface finish and corrosion resistance assessment are now well defined, the other four characteristics in the top six still do not have recognised standard test methods. Our work has shown that the following methods are highly suitable for the properties numbered 1, 2, 5 and 6:

Scratch test
Ball crater
Pin-on-disc
Vickers microhardness

However, these must be used with extreme caution and the interpretation of the results must recognise the scatter which is inevitable in readings taken by all these methods, even on nominally identical coatings.

ACKNOWLEDGEMENTS

We thank the EC, EPSRC and industrial sources for financial support. Special thanks go to our collaborators at the VTT and THD laboratories, and to our colleagues in the RCSE for help and advice.

REFERENCES

1. H. Ronkainen, S. Varjus, K. Holmberg, K. S. Fancey, A. R. Pace, A. Matthews, B. Matthes and E. Broszeit: *Proc. Leeds–Lyon Conference, Lyon, France, 1989*, Tribology Series, Vol. 17, D. Dowson, C. M. Taylor and M. Godet eds, Elsevier Sequoia, 1990, 453.
2. B. Ollivier and A. Matthews: submitted to *J. Adh. Sci. and Technology.*

3. J. Valli, J. Palojarvi and U. Makela: *Research Notes 435*, Technical Research Centre of Finland, 1985, 22.
4. A. E. Almond and M. G. Gee: *Wear*, 1987, **120**, 101.
5. H. Czichos, S. Becker and J. Lexow: *Wear*, 1987, **114**, 100.
6. S. Bennett, A. Matthews, J. Valli, A. J. Perry, S. J. Bull and W. D. Sproul: *Presented at ICMCTF '92, San Diego, USA*, published in *Tribologia*, (The Finnish Journal of Tribology), 1994, **13**, 16–28.

WEAR TESTING OF COATINGS: METHODS, SELECTION AND LIMITATIONS

C. SUBRAMANIAN
*Surface Engineering Research Group, Gartrell School,
University of South Australia, The Levels, SA 5095, Australia*

ABSTRACT

Appropriate wear testing is a prerequisite for a material to be considered for possible tribological applications. Engineers and designers need wear rate data to confidently design components and tools with reliability and quality. The push for improved performance of engineering artefacts has driven the developments of new materials and coatings. Before these new materials and coatings can be used, they should be characterised and evaluated for tribological performance.

Tribological evaluation, in particular wear testing, can be performed in many ways. There are numerous testing machines available from a simple laboratory type (e.g. pin-on-disc) to fully simulated rigs (e.g. engine testing). This paper describes various types of wear, reasons for testing, selection of wear testing methods and their limitations. The applications of appropriate wear testing methods to thin coatings and surface treatments in particular are discussed. The problems associated with the terminology and units used in the measurement of wear are also highlighted. The need for a standardised approach to wear testing of coatings is emphasised.

1. INTRODUCTION

There is a growing interest in extending the quality and reliability of artefacts in mechanical systems. This has led to the development of new materials and coatings which can withstand the severe conditions imposed by designers and engineers. If a monolithic material is unsuitable for a particular application, composite materials and structures are actively considered. Such composite structures include coatings and surface-modified materials. Generally these surface engineered systems have a soft, tough core and a hard, mostly brittle, outer layer. However, soft coatings on hard materials are not uncommon, e.g. silver coatings on steel where silver acts as a solid lubricant. Modification of material surfaces is both old and new – it is new because there is an ever increasing number of techniques being developed to coat or treat a material surface; it is old as can be seen from the history of heat treatment – surface hardening of steel has been in practice for many centuries.

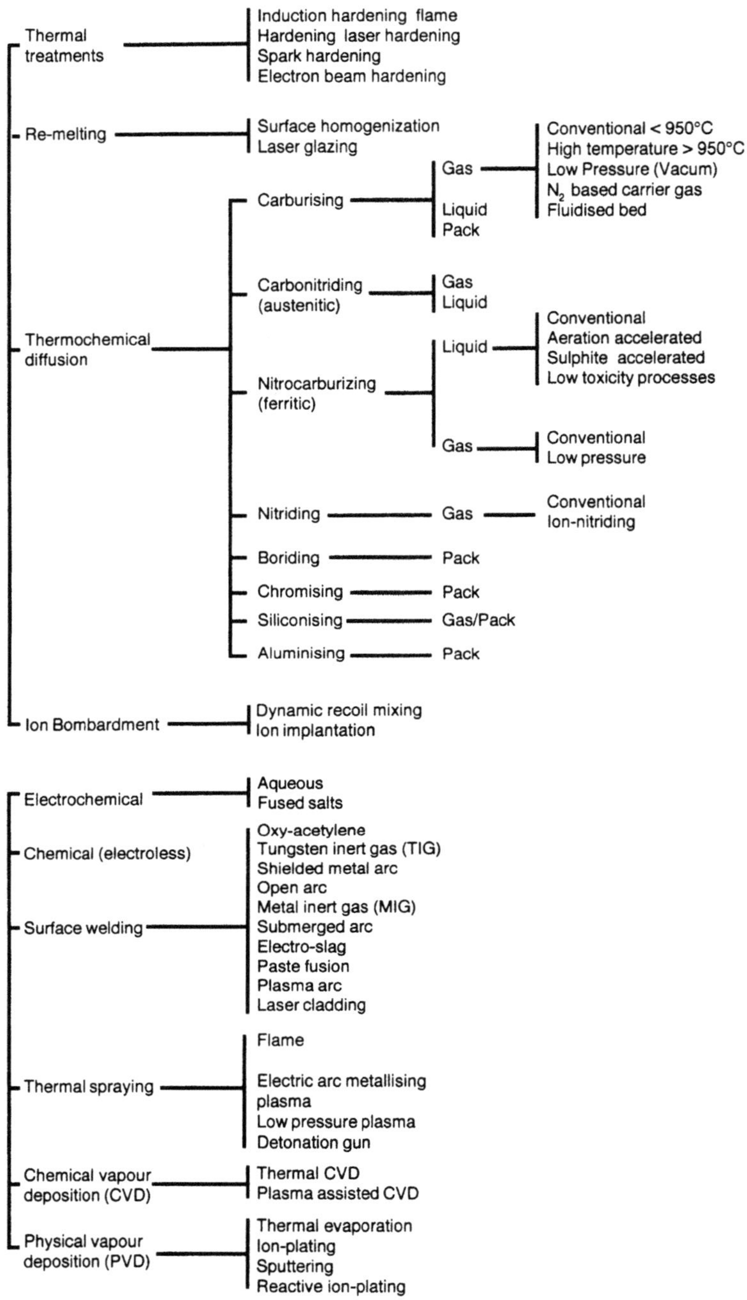

Fig. 1. Various methods of surface coatings and treatments.

1.1 Surface Engineering

Surface Engineering has been defined as 'the application of traditional and innovative surface technologies to engineering components and materials in order to produce a composite material with properties unattainable in either the base or surface material.'[1] One of the main aims of surface engineering is to improve properties such as friction, wear, corrosion, fatigue and bio-compatibility. The basic idea of surface engineering is to use leaner (often cheaper) materials as substrates and rare or strategically important (often expensive) materials as coatings.

The traditional methods such as carburising, flame hardening and shot peening and the newer ones such as physical vapour deposition (PVD), chemical vapour deposition (CVD) and laser processing, have been successfully employed to improve the performance of many engineering components in service. These methods are listed in Fig. 1. The thicknesses of various coatings or the depths of surface-modified material are compared in Fig. 2.[1] It is evident from Fig. 2 that the layer could be as

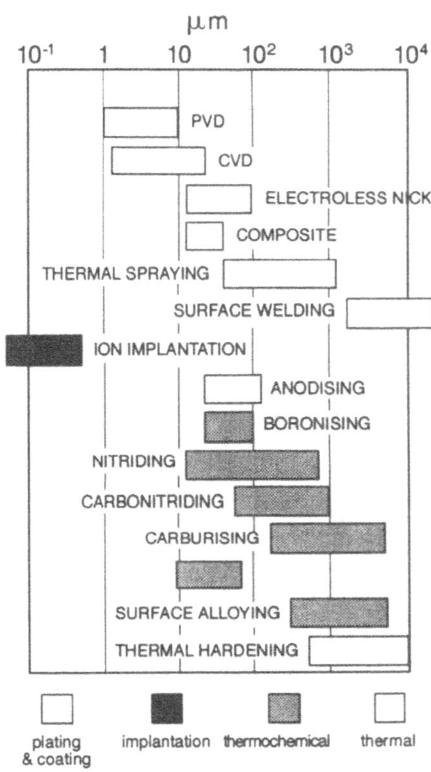

Fig. 2. Thickness of various surface coatings and treatments.

thin as few atomic layers in ion-implantation, to as thick as a few mm in flame sprayed/weld-deposited coatings.

1.2 Tribology

A standard definition for tribology is 'the science and technology concerned with interacting surfaces in relative motion'.[2] It originates from the Greek word *tribo* which means 'rubbing'. It embraces the study of friction, wear and lubrication. Tribology is truly an interdisciplinary area of research which is of interest to physicists, chemists, materials scientists and engineers. *Friction* has been a main concern of physicists, whereas *lubrication* has largely been in the research domain of chemists. The third aspect of tribology, *wear*, has traditionally been dealt with by mechanical engineers. However, in the last 30 years or so the *systematic* study of wear has attracted materials scientists and metallurgists. This might perhaps be due partly to the inability of the mechanical engineer to 'see' beyond the continuum concepts of materials and to the inherent ability of the metallurgist to recognise the importance of microstructure and structure/property relationships in general. The material scientists' contribution to wear has been very significant as it is evident from the published literature concerning the role of microstructure on the tribological properties of materials.

Wear is a complex phenomenon as many factors influence it. A system approach has been suggested to study wear and its related constituent elements.[3] A typical wear system is shown in Fig. 3 with input and output parameters. Considering the

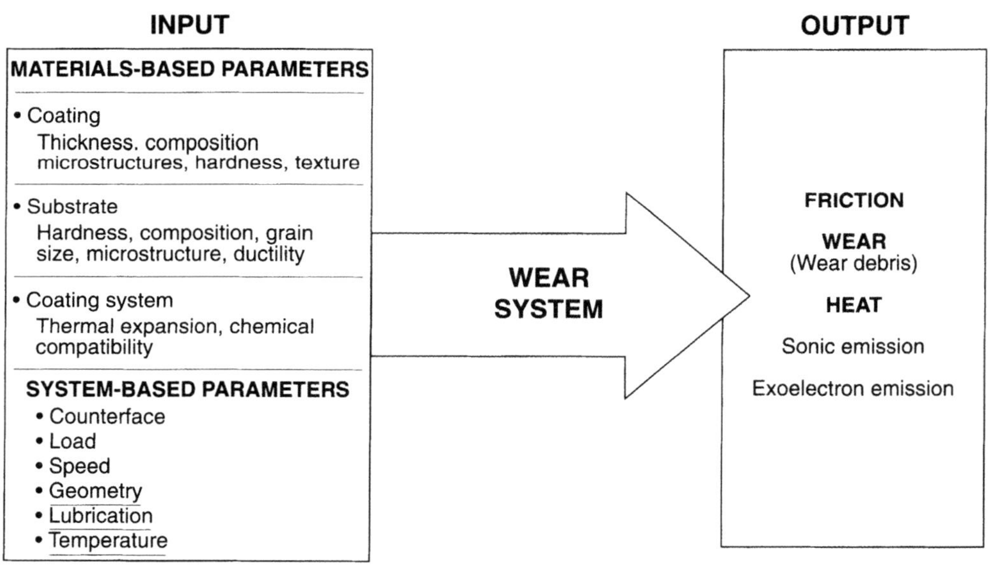

Fig. 3. A wear system with input and output parameters.

number of variables in the system, it is not surprising that wear data collected under well-controlled laboratory conditions cannot be directly correlated to those obtained in the actual working environment. Another reason for poor correlation between laboratory and field test data is the common concurrent occurrence of two or more wear mechanisms in the system under actual conditions of application.

There is no formal classification of wear which has been fully accepted by researchers in the field of wear. However, to deal with a complicated problem such as wear, we need some classification based on one or more criteria. One of the most widely used wear classifications is due to Burwell and Strang[4] which is mainly based on four types of wear: abrasion, adhesion, corrosion and fatigue. Although these four types appear to uniquely represent various wear phenomena, there are a few drawbacks:

(i) it is not a complete description;
(ii) it does not describe the production of wear particles from the surface and eventual ejection out of the wear system;
(iii) two or more mechanisms of debris production may operate simultaneously or successively.

A recently developed classification of wear is due to Lancaster[5] who advocates material-specific wear types: metals, ceramics and polymers. These and other classifications are listed in Table 1. It is therefore advisable to mention the actual

Table 1. Classifications of wear.

Basis of classification	Wear types (e.g.)
1. General	Adhesive, abrasive, corrosive, fatigue
2. Motion	Sliding, reciprocating, impacting
3. Geometry	Pin-on-disc, pin-on-ring, cylinder-on-cylinder
4. Mechanism	Delamination, oxidation, seizure, adhesion
5. Applied load	Low stress, high stress
6. Lubrication	Lubricated (boundary, mixed, hydrohynamic) unlubricated (dry)
7. Material	Metal, ceramic, polymer, composite

condition by stating the type of geometry, lubrication, materials involved and the predominant mode(s) of debris production, rather than just stating an assumed wear mechanism (e.g. adhesion, delamination, etc.). Any wear study should concentrate on the nature of the worn surfaces and the wear debris in order to shed some light on the wear mechanism(s) of debris particle formation and the eventual release out of the wear system. If the wear particles formed in a wear system are not removed, they will lead to different wear mechanisms such as the three-body situation which has been extensively studied by Codet and his workers.[6]

2. WEAR TESTING

Wear tests are carried out for various reasons:

(a) to evaluate the resistance of a material or a pair of materials to wear;

(b) to evaluate the performance of lubricants;
(c) to determine the friction coefficient between two materials;
(d) to understand the fundamentals of tribology (generally under well-controlled conditions between pure metals);
(e) to estimate the lifetimes of components and tools used in engineering and other applications.

Table 2 lists various ASTM standards for wear testing.[7] The methodology of wear testing has also been set out in published literature, notably in the Special Technical Publications by ASTM.[8–12] Among the ASTM books on wear testing, there is one for coatings.[10] This book deals with many coatings such as plastic coatings, electroless nickel coatings, etc. However, it fails to discuss the special problems encountered with wear of hard thin films.

Wear testing can be done in many ways: from a 'simple' test in a laboratory to a determination in the actual field. The cost involved in a laboratory test is very small when compared to a test in the actual environment. However, laboratory test data may not be useful unless compared and correlated with those obtained in the field tests. Often, it is found that these sets of data are far apart, unless due care is taken in the laboratory to attempt to simulate the actual field conditions. Generally, the load and speed levels used in the actual condition are adopted in the laboratory test. Other variables such as the geometry, the counterface and the size of the specimens may only be occasionally duplicated in the laboratory. When everything is duplicated, the laboratory machine will be a copy of the actual rig! It is worth considering various stages of wear testing.[13] The appropriateness of the level of simulation of the wear test can be selected depending on the element of risk and the closeness to the actual application desired.

There has been many studies conducted on the wear behaviour of coatings and surface-modified materials in laboratory testing machines and actual field conditions. However, there are only a few investigations in the literature which attempts to carefully relate these sets of wear data observed in laboratory to the field data.

3. WEAR TESTING OF COATINGS

As mentioned before, the best method for testing a coating for its wear behaviour is to put it into service. As this is expensive, time-consuming and sometimes catastrophic, some testing facilities have to be devised which are simple and easy to use and, at the same time, relevant to the application of the particular component. Table 3 shows a list of selected wear test devices most commonly used (data from many published papers).

Depending on the thickness of the coating, the wear test methodology should be chosen. Thicker coatings (ca. $>10\,\mu m$) should be treated as bulk monolithic materials, whereas for thin coatings (ca. $<10\,\mu m$) different criteria should be applied. In the following sections, five aspects will be considered:

Table 2. ASTM wear tests.

Materials*	Comments	ASTM Code†	
cemented carbides	slurry abrasion, flat against wheel	B 611	M
mineral aggregates	crushing resist., Los Angeles mach., ball mill	C 131	M
concrete	air-driven sand blasting	C 418	M
Cer porcel. enamel	Nat. Bur. of Standards lapping machine	C 448	M
Cer tile	Taber abraser	C 501	M
mineral aggregates	crushing resistance (see C 131)	C 585	M
Cer refract. brick	room temperature jet erosion	C 704	M
concrete	3 machines: abrasion, sliding, impact	C 779	M
C-graphite	reporting data in seal wear tests	C 808	G
concrete	rotary cutter on a drill press	C 944	M
organic Ctg	jet-abraded coupons of paint, lacquer, varnish	D 658	M
organic Ctg	falling sand/SiC on paint or lacquer coupons	D 968	M
organic Ctg	wear of floor coverings – see D 658 and D 968	D 1395	M
rubber, leather	shoe soles and heels, abrasive drum machine	D 1630	M
Met, Cer, Pls	ring-on-block machine	D 2714	M
Tex	clothing wear on human subjects	D 3181	M
Met	Falex #6 thrust washer test for self-lubr. Met	D 3702	M
Tex	Taber abraser for abrasion resistance	D 3884	M
Tex	rub while flexing woven fabrics	D 3885	M
Tex	rub woven fabrics on an inflated diaphragm	D 3886	M
Tex	oscillating cylinder abrasion	D 4157	M
Tex	rotary rubbing abrasion (Shiefer machine)	D 4158	M
organic Ctg	floor covering wear using the Taber abraser	F 510	M
insulation	revolving drum test for pipeline Ctg	G 6	M
Met, Cer, Pls	vibrating cavitation horn test	G 32	M
Tex	wear of inked ribbons, ribbon-wrapped drum	G 56	M
Met	dry flowing sand, rubber rotating wheel	G 65	P
Met, Pls	liquid erosion by jets, droplets, spinning rig	G 73	P
Met	slurry abrasivity, immersed reciprocating lap	G 75	M
Met, Cer, Pls	erosion, particles impinging on a flat coupon	G 76	P
Met, Cer, Pls	sliding wear, block-on-ring – see D 2714	G 77	P
Met, composites	jaw crusher plate wear simulator	G 81	P
Met	crossed cylinders	G 83	M
greases	four-ball machine, anti-wear properties	D 2266	M
greases	block-on-ring Timken test, E.P. properties	D 2509	M
greases	four-ball machine, E.P. properties	D 2596	M
greases	oscillating block-on-ring, anti-wear properties	D 3704	M
greases	Fafnir friction oxidation test, fretting	D 4170	M
dry lubricants	adhesion of tape on anodized Al panels	D 2510	M
dry lubricants	heat and quench steel panel (thermal shock)	D 2511	M
dry lubricants	steel pin in V-block, load carrying capacity	D 2625	M
dry lubricants	ring-on-block	D 2981	M
fluid lubricants	load-carrying capacity of gears, Ryder machine	D 1947	M
fluid lubricants	pin in V-block, seizure test	D 2670	M
fluid lubricants	lpin in V-block, two methods for E.P. prop.	D 3233	M
fluid lubricants	anti-wear, four-ball machine	D 4172	M
hydraulic fluids	wear of parts in recirculating pumps	D 2271	P
hydraulic fluids	wear of parts in recirculating pumps	D 2882	M

* Met = metals/alloys, Cer = ceramics, Pls = plastics, Ctg = coatings, Tex = textiles.
† M = test method, G = guideline, P = practice, * = discontinued.

Table 3. Various wear testing methods needed for tribological evaluation of different surface treated/coated materials (data from published sources).

Wear testing machine	Conditions	Materials used
1. Pin-on-disc sliding wear tester	0.1–300 N load 0.0016–3 ms^{-1} speed	Ion implanted, CVD, PVD electroless coating, electroplating, laser processed, plasma sprayed
2. Abrasion testers (a) Dry sand Rubber Wheel	30–250 N load 195 rpm speed	Laser processed, Cr plated plasma processed
(b) Taber Abrasion	10 N load 60 rpm speed	Electroless plating, PVD
(c) Polishing wear	6–10 KPa pressure 0.23 ms^{-1} speed	Ion-implanted, PVD, CVD thin coatings
3. Erosion	150–200 ms^{-1} speed	Plasma sprayed, PVD, CVD
4. Roll-on-roll	49–2000 N load 250–3000 rpm speed	PVD, CVD, laser-treated, diffusion treated, plasma nitrided
5. Pin-on-ring	0.5–100 N load 0.002–0.8 ms^{-1} speed	PVD, electroless coating
6. Crossed cylinders	90–300 N load 11–100 rpm speed	Ion-implanted, electrospark deposited

(a) selection of wear testing machine;
(b) selection of testing parameters;
(c) wear rate measurement methods/criteria and post-wear analysis;
(d) field tests – metal cutting tools, forming tools;
(e) comparison of wear data – laboratory and field tests.

3.1 Selection of Wear Testing Machine

The American Society of Lubrication Engineers (ASLE) (presently the Society of Tribologists Lubrication Engineers – STLE) published a list of more than 200 different types of wear testing machines in the middle of the 1970s.[14] The important test geometries are shown in Fig. 4. The pin-on-disc machine has been considered to be a *de facto* standard.[15] Two other popular machines are pin-on-ring and reciprocating type rigs. For coatings, generally the flat pin is replaced by a ball or hemispherical pin which will facilitate the seating of the pin or ball without much trouble on the counterface. These machines are generally used to simulate sliding wear situations.

For abrasive wear testing there are many types of machines available.[16] Some of them have been used for evaluating abrasive wear resistance of coatings.[17] Rabinowicz's wear testing technique, using a lapping machine,[18] was modified by Singer and his co-workers,[19,20] which gives a depth resolution of 10 nm. Similar to depth profiling of an element obtained by Auger analysis, this technique provides wear rate/resistance as a function of depth from the surface. For example, a wear resistance profile was matched to N^+ concentration profile in an N^+ ion-implanted carbon steel.[19]

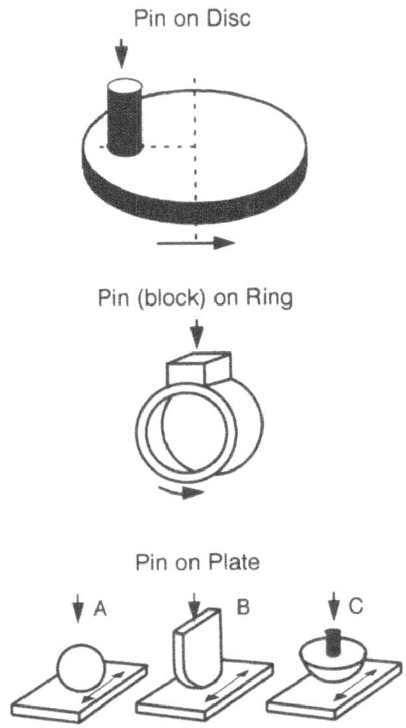

Fig. 4. Most widely used wear-testing geometries.

A few problems arise with softer specimens (<400 HV) in the presence of very small abrasive particles. However, this method is suggested to be very useful in estimating the abrasive wear resistance of very thin layers such as for ion-implanted materials and very thin PVD coatings. For example, this method has been successfully used to evaluate the wear behaviour of ion-implanted 304 stainless steel, 52100 steel and titanium alloys, coatings produced by PVD and CVD methods and laser-alloyed surface composites.

The selection of the wear testing machine depends on many factors such as the closeness of the service condition to that simulated by the testing machine, availability, etc. In many instances, the availability of the wear testing determines the type of the machine to be used. This is because a commercially available machine cost typically $10,000–$100,000 or more, depending upon the sophistication involved. Generally 'home-made' ones cost a lot less. The lack of funding for tribology studies often forces researchers to use the available testing machines though it may be partly inappropriate.

The type of the testing machine or machines to be selected depends on the modes of wear to be encountered in the actual application. For example, if the components in service failed by a mixed mechanism of adhesive and abrasive (3-body low stress) wear mechanisms, it is prudent to select two machines which can simulate these mechanisms. Then there is a possibility that the test data obtained from these two tests will mean something in the actual application.

3.2 Selection of Testing Parameters

Important parameters such as applied load, relative speed, countersurface, relative humidity and operating temperature should be considered before the commencement of wear testing.

3.2.1 Load

It is generally observed that wear increases with applied load. Accelerated wear tests using higher loads could lead to different wear mechanism and thus higher or lower wear rates than that expected in the real life situation. Wear maps[21,22] should be used as a broad guide in determining possible wear mechanisms and resultant wear rates. An apparent pressure which is close to the real-life situation should be applied in any simulated test.

3.2.2 Velocity

The relative speed of the contacting bodies is as influential as the applied load – this can be seen from the fact that velocity is one of the parameters used in the construction of the wear maps. It also directly influences the temperature developed at the sliding interface. As far as possible, in laboratory testing, velocity of the tribocouple should also be realistically simulated in order to obtain the same wear mechanism as in real-life situation.

3.2.3 Specimens

Having made the decision on the type of the testing rig, the next question is which part is chosen to be the specimen – the pin or the disc? This is an important consideration because some studies have proved that there is a significant difference in wear rates associated with reversing the pin and disc roles[23] and by changing the counterface.[24] Also considered important to be kept constant throughout the testing programme are the roughness, microstructure, homogeneity of composition and preparation and cleaning schedule of the specimens.

3.2.4 Others

Other important parameters which should be monitored and controlled include relative humidity and temperature of the laboratory.

3.3 Wear Rate Measurement

The wear of the coating can be measured in many ways – volume or mass loss, dimensional change, etc. Depending on the density of the coating material, sometimes the weight loss method results may be misleading, especially when comparing two coatings with differing densities. In such situations, volume loss or dimensional change would be appropriate. Other wear measuring/monitoring techniques include visual or tactile inspection, profilometry, radiography, spectroscopy of wear particles and acoustic (noise) signal measurement.

These wear volumes can be expressed as a wear rate which is wear volume/unit sliding distance or time. The thickness of the coating and the sliding speed are important here, as these two will determine the duration of the test. As mentioned in the Metal Handbook, the most appropriate test duration for a thin coating is that continued until the coating wears out.[25] Thicker coatings obviously can be tested for longer time before the wearing surface reaches the coating–substrate interface. In many instances the commencement of interaction between the wearing interface and the substrate is hard to judge. However, in some cases, the change in colour of the coating, friction co-efficient, or noise could be an indication that the coating is worn out.

Measurement of wear volume or change in dimensions may be continuously monitored or periodically measured. In these ways the history of the coating failure in wear testing will be known. Along with wear, friction co-efficient is also generally recorded.

Generally wear by mass loss is measured by sensitive balance (accuracy 0.1 mg). This determines the maximum weight of the specimen which can be weighed with the required accuracy (normally below 65 g). The heavier the specimen, the less accurate the reading. If heavier specimens are to be used, surface profilometry can be employed to measure the wear volume from the profile of the wear scar. There are contact and non-contact type surface profilometers with 2D or 3D mapping available commercially. The width, or more generally, the wear depth of the wear scar, may be used as a criterion of wear.

3.3.1 Post-wear Analysis

There is a gamut of analytical techniques available for examining the wear debris particles and the worn surfaces. Among them, the most commonly used techniques are scanning electron microscopy (SEM) of wear particles and worn surfaces for characterising the morphology of debris and the surface appearance of the worn surfaces. SEM combined with analytical facilities is very useful in characterising for wear mechanism(s). X-ray diffraction is used to determine the phase(s) present in debris and sometimes on the worn surface. Space does not permit discussion of all the techniques here, but details can be found elsewhere.[26,27]

3.4 Field Tests

Depending upon the application of the coatings, some field tests should be undertaken in order to validate the laboratory data. Also, the field test is the 'best' test for a coating or surface treatment where it is meant to be used.

The PVD/CVD coatings have been developed generally for applying in metal-cutting tools, and metal-forming dies and tools. Therefore, the tests reported in the literature to evaluate the behaviour of the coatings commonly involve actual testing of coated parts in actual service situations. The original uncoated substrate is used as a control specimen.

3.4.1 *Machining Tests*

Direct and indirect measurements of various tool wear criteria have been reported.[28] One direct method of measuring tool wear is to stop the machining operation periodically and record the amount of wear (crater depth, wear on the clearance face, etc.) either using optical or electron microscopes or via special instruments which measure the topographic features of tools. Indirect measurements of wear of cutting tools include measurements of cutting force, cutting temperature, tool deflection, tool vibration, power input to the machine tool, autoradiography of wear particles and electrical resistance between the tool and the workpiece.[29–32] These indirect measurement techniques are vitally important as a new era of machining enters complete automation. However, it should be mentioned here that more work needs to be conducted in correlating such indirect measurements of tool wear with the actual wear behaviour.

The running of a machine test includes selection of various parameters involved in machining such as machine tool to be used, tool geometry, feed rate, the depth of cut, lubrication, workpiece material speed etc. Generally, progressive tool wear is measured for various speeds, keeping all other parameters constant. If a Taylor tool life plot of log tool life vs. log speed is established for various tools, then the performance of tools may be compared.[29]

3.4.2 *Tests for Metal-forming Dies and Tools*

There is a growing interest in the application of surface engineering solutions to the wear problems of tools and dies used in metal forming processes.[33] Generally, dies and tools used in metal-forming operations experience elastic deformation whereas workpieces undergo plastic deformation. This is starkly different from the metal-cutting situation, where there is severe deformation of the workpiece and the tool coupled with the development of high temperatures at the chip–tool interface. This is one of the reasons why certain surface treatments (e.g. ion implantation) perform very well under metal-forming operations but fail in metal-cutting situations.

A scan through the literature indicates that any surface treatment or coating applied to metal-forming dies and tools has largely been tested under actual application conditions,[34] though, in some cases, simulation of such operations has been performed.[35] Wear in such dies and tools involves adhesion (metal pickup), abrasion, fatigue (spalling) and corrosive wear (for functioning in a corrosive environment). As mentioned previously, the simultaneous occurrence of two or more such basic wear modes makes any simple laboratory test inappropriate.

When adhesion, hardness and other basic characterisation of a coating are evaluated, it is suggested that coating be tested in actual metal-forming operations. Periodic monitoring of wear of dies/tools or the workpiece is by visual inspection for obvious pick up, and by profilometry and microscopy.

3.5 Correlation between Laboratory and Field Test Data

As mentioned before, wear data obtained in a laboratory wear testing machine need not represent the true performance in actual field applications. This is because of differences in specimen geometry, size, type of environment, load, speed, etc. Depending on load or speed, wear mechanism often changes from one to another. These aspects have been discussed at length in papers dealing with wear mechanism maps.[21,22]

Wear mechanism maps displaying wear rate contours are drawn using dimensions of pressure and velocity as axes. These maps can be used to predict wear mechanisms and possibly wear rates under certain conditions. They can also be applied to actual field conditions such as in metal cutting, Fig. 5.[36]

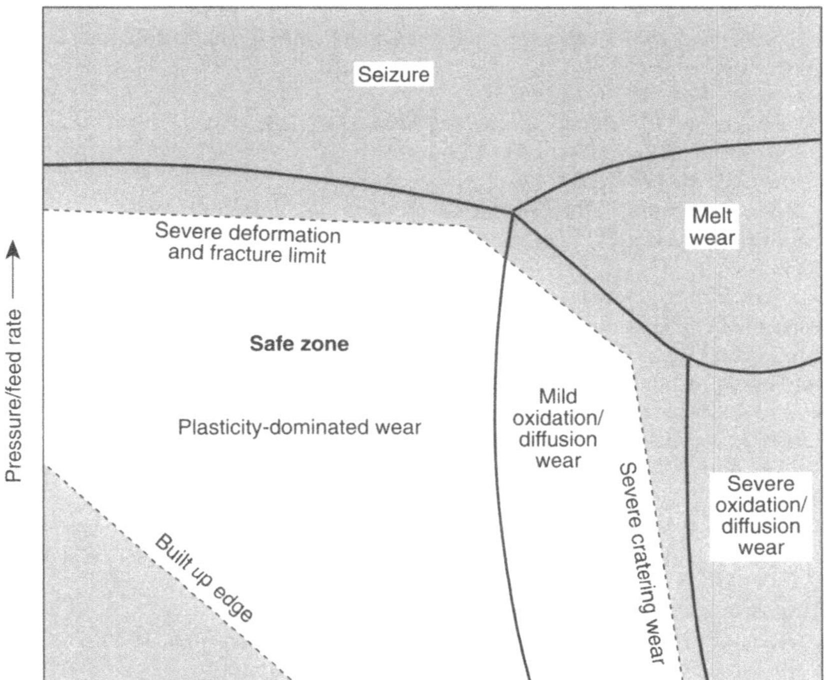

Fig. 5. Application of wear mechanism map to metal cutting situation.

4. CONCLUSIONS AND RECOMMENDATIONS

The wear testing of coatings and surface-modified materials are essentially similar to those used for bulk materials except the fact that the sensitivity of the measurement of wear rate should be higher in the former. The large variations in published results of the wear rates of coatings may be due to overlooking the thickness factor and the variation of properties (microstructure, composition, etc.) from the surface towards the bulk. Wear mechanism maps for coatings with parameters, which are relevant to the real-life situations, should be developed to enable better prediction of wear mechanisms and wear rates and possible correlations between laboratory and field data. Standardised test geometry and testing procedure should be developed for thin and thick coatings.

REFERENCES

1. T. Bell: *Metals Mater.*, 1991, **7**, 478.
2. H. P. Jost: *Lubrication (Tribology) Education and Research*, The Jost Report, London, H.M. Stationery Office, 1966.
3. H. Czichos: *Wear*, 1977, **41**, 45.
4. J. T. Burwell and C. D. Strang: *J. Appl. Phys.*, 1952, **23**, 18.
5. J. K. Lancaster: *Wear*, 1991, **141**, 159.
6. M. Godet: *Wear*, 1984, **100**, 437.
7. P. J. Blau: *Friction and Wear Transitions of Materials*, Noyes, NJ, 1989.
8. 'The Selection and Use of Wear Tests for Metals', R. G. Bayer, ed., ASTM Special Tech. Publ. 615, 1976.
9. 'Wear Tests for Plastics: Selection and Use', *ASTM Special Tech. Publ.* 701, 1979.
10. 'Selection of Wear Tests for Coatings', *ASTM Special Tech. Publ.* 769, 1982.
11. 'Selection and Use of Wear Tests for Ceramics', *ASTM Special Tech. Publ.* 1010, 1988.
12. 'Wear Testing of Advanced Materials', *ASTM Special Tech. Publ.* 1167, 1992.
13. H. Uetz, K. Sommer and M. A. Khosrawi: *VDI Berichte*, 1979, **354**, 107.
14. *ASLE Friction and Wear Devices*, 2nd ed., 1976.
15. T. S. Eyre and F. A. Davies: *Surface Stability*, T. N. Rhys-Jones, ed., The Inst. of Metals, 1989, 186.
16. R. Blickensderf and G. J. Laird II: *Testing and Evaluation*, 1988, **16**, 516.
17. S. J. Bull, D. S. Rickerby, T. Robertson and A. Hendry: *Surf. Coat. Technol.*, 1988, **36**, 743.
18. E. Rabinowicz: *Lubrication Engineer*, 1977, **33**(7), 378.
19. I. L. Singer, R. N. Bolster and C. A. Carosella: *Thin Solid Films*, 1980, **73**, 283.
20. R. N. Bolster and I. L. Singer: *ASLE Trans.*, 1981, **24**, 526.
21. S. C. Lim and M. F. Ashby: *Acta Metall.*, 1987, **35**, 1.
22. R. Antoniou and C. Subramanian: *Scripta Metall.*, 1988, **33**, 809.
23. S. L. Rice, S. F. Wayne and H. Nowotny: *Wear*, 1983, **88**, 85.
24. C. Subramanian: *Scripta Metall. Mater.*, 1991, **25**, 1369.
25. *ASM Metals Handbook*, 9th edition, vol. 8, *Mechanical Testing*, 1985, 601.
26. E. R. Weber: 'Auger Electron Spectroscopy for Thin Film Analysis', *Res. Dev. Magazine*, 1972, **23**(10), 22.
27. D. H. Buckley: *Surface Effects in Adhesion, Friction, Wear and Lubrication*, Tribology Series 5, Elsevier, 1981.
28. V. A. Tipnis: *Wear Control Handbook*, M. B. Peterson and W. O. Winer, eds, ASME, NY, 1980, 891.
29. E. M. Trent: *Metal Cutting*, 2nd ed., Butterworths, 1984.

30. M. C. Shaw: *Metal Cutting Principles*, 3rd ed., MIT Press, MA, 1954.
31. S. Kalpakjian: *Manufacturing Processes for Engineering Materials*, Addison Wesley, MA, 1984.
32. J. A. Schey: *Tribology in Metalworking – Friction, Lubrication and Wear*, ASM, OH, 1983.
33. C. Subramanian and K. N. Strafford: *Materials and Design*, 1993, **14**, 291.
34. G. Dearnley: *Proc. Inst. Mech. Engr.*, 1980, 15.
35. J. Halling: *Proc. Inst. Mech. Engr.*, 1980, 29.
36. L. A. Kendall: *Metals Handbook*, 9th ed., vol. 16, Machining, 1989, 37.

ADVANCED SURFACE ANALYTICAL TECHNIQUES IN CHARACTERISATION AND QUALITY ASSURANCE OF COATINGS

R. St. C. SMART, A. K. O. NETTING, W. M. SKINNER, M. FIELDEN, R. A. HAYES and J. RALSTON

Ian Wark Research Institute,
University of South Australia, The Levels, SA 5095, Australia

ABSTRACT

The characterisation and quality assurance of coatings in surface engineering of materials is critical to both development and process control in diverse areas of application. The availability of advanced surface analytical techniques for measurement of chemical and structural properties of these surfaces has markedly increased over the last decade. The increasing sophistication of the technologies for production of thin films and coatings has, however, not been matched in general by similar increases in sophistication of the measurements related to characterisation and QA of these surface layers. The type of information relating to surface properties and features required to provide understanding of processes (and failures) is summarised for a variety of materials, thin films and coatings. The depth of analysis and the spatial resolution (from atomic detail to mm) to meet these requirements are considered. Composition (stoichiometry), impurities, depth profiles, topography, defects, phase structures, interfacial bonding and surface reaction comprise some of this information.

Eight techniques for obtaining this information are reviewed, namely: X-ray photoelectron spectroscopy; scanning Auger microscopy/spectroscopy; secondary ion mass spectrometry; analytical scanning electron microscopy; analytical transmission electron microscopy; Fourier transform infrared spectroscopy; scanning tunnelling microscopy/ spectroscopy; and atomic force microscopy. The basis of each technique is briefly reviewed together with the type of information and limitations in measurements. The use of STM and AFM, in particular, to provide structural detail down to the atomic level complements the chemical information available from XPS and SAM.

Case studies of the applications of XPS, SAM, AFM and STM to studies of film and coatings systems are given in the last section. A coating system of electroless nickel over titanium nitride on a mild steel substrate is characterised using SAM. XPS studies of oxidation in the columnar structure of a titanium nitride PVD-coated mild steel are described. The information available from AFM studies of silane-modified oxidised silicon is reviewed for topographic information, force–distance measurements and surface roughness quantification. The oxidation of a sulphide surface in air and solution is followed systematically using STM relating the corrosion process to surface defect and impurity sites.

1. INTRODUCTION

The characterisation and quality assurance (QA) of coatings in surface engineering of materials lies at the heart of both the development and process control in diverse areas of application. Wear resistance, corrosion resistance, hydrophobicity, chemical surface modification for incorporation into composite matrices, electrical or thermal insulation and reflectivity/surface finish engineering are just some of the application areas in which characterisation and QA of surface layers are required. There are many time-honoured, reliable methods for testing the performance qualities of thin films and coatings based on mechanical and imaging techniques. Some of these techniques will be reviewed elsewhere in this volume. Increasingly, however, there is demand for more specific compositional and structural information on these surface layers at levels of detail down to the atomic and molecular scale. These demands have been generated by the increasing sophistication of the technology for the production of the thin films and surface coatings including physical vapour deposition (PVD), chemical vapour deposition (CVD), plasma treatments, Langmuir–Blodgett (LB) films, adsorbed layers, electrodeposited layers and thermal treatments. The field of surface engineering, using these sophisticated technologies, has emerged as one of the most rapidly growing areas of research, development and processing in the last two decades. The requirements for control of surface properties are now of paramount importance in most of the new manufacturing and 'high' technologies. The selection of materials surfaces to meet increasingly demanding applications, with the concomitant requirements for QA of the surface properties, is now a recognised part of the economics of most of these industries. Costs associated with high-throughput processes, increasing chemical and mechanical stresses, materials lifetimes and safety now place major demands on surface engineering solutions.

The increasing sophistication of the technologies for production of thin films and coatings has not, in general, been matched by similar increases in sophistication of measurements related to characterisation and QA of these surface layers. Most of the routine analysis in these areas is still carried out using metallurgical and chemical techniques originally developed for assessment of metals, alloys, ceramics and polymers. This is not intended to imply that these methods are not valuable or that they should be discarded or modified. Most of this testing will continue to be required with the new surface engineered coatings to provide mechanical and structural information together with chemical information on the *bulk* properties of the materials. The advent of new surface analytical techniques over the last decade now offers the possibility of very detailed information on the surface layers and profiles through deposited thin films and coatings. It is now possible to very fully define the compositional and structural properties of these films across the millimetre to sub-nanometer levels of detail. Some of these techniques are sufficiently time-consuming that they are best suited to problem solving and diagnostic applications in the development phase or to cases where process control has been lost or compromised (e.g. scanning tunnelling microscopy, atomic force microscopy). Other advanced surface analytical techniques, however, are now increasingly being used in routine QA applied to thin film technology, coatings and surface modification processes (e.g. analytical scanning electron microscopy, X-ray photoelectron spectroscopy, scanning Auger microscopy/spectroscopy, Fourier transform infrared spectroscopy and imaging secondary ion mass spectrometry).[1] Industrial laboratories in many countries have now installed these techniques for batch monitoring as well as problem solving and diagnostic applications.

In this short review, we will set out the type of information generally sought in characterisation and QA of coatings. This information can be divided into categories of essentially compositional and chemical state measurements as opposed to essentially structural and topographic information. In the second section, eight surface analytical techniques will be briefly surveyed with emphasis on the new types of information available from atomic force and scanning tunnelling microscopies used in different imaging and spectroscopy modes. Finally, a number of case studies applying some of these techniques to surface coatings will be summarised including titanium nitride, silane-modified silica surfaces and oxidation processes at sulphide surfaces.

2. SURFACE INFORMATION

The importance of material surfaces in technological application is now difficult to overstate since they directly influence so many properties in materials application. The control of material stability via corrosion and frictional wear, material adhesion characteristics, ceramic binding, electronic device fabrication, membrane structure and functions, catalysis, composite formation and strengths, and polymer product formation are all major facets of modern industry. The information required to understand the processes and final products in detail may dictate the use of a variety of different techniques. In industrial process control, when problems arise they usually require immediate solutions. Knowledge of the factors and mechanisms involved in the formation of the films and coatings is then essential if time-consuming empirical testing and guesswork are to be avoided. The type of information relating to surface properties and features that will provide this level of understanding is summarised for a variety of materials together with thin films and coatings in Table 1.

The depth from which surface information needs to be collected varies considerably between one material and another and between one application and another. For instance, Table 2 indicates the importance of various different depth regimes related to surface functions in application.[1] Analysis depths ranging from a single molecular or atomic monolayer through to layer thicknesses of tens of micrometers, even millimetres, may be required to fully describe the composition

Table 1.

Material	Surface features
Thin films and coatings (PVD, CVD, adsorbed layers, LB film, paints, passivation layers, electrodeposition etc.)	Composition (stoichiometry), impurities, depth profile, structure (topography, defects, phases, grain sizes, monolayer/multilayer order, coverage, continuity); oxidation and reaction (corrosion, weathering); interfacial bonding (layer/substrate); inter diffusion (layer/substrate)
Metals, alloys	Oxide coating (thickness, type); faceting; relaxation; reconstruction, elemental segregation; defects; fracture faces; corrosion layers and regions; adsorbed layers; interdiffusion and reaction (joining technology)
Semiconductors (including organic semiconductors)	Space-charge region; reconstruction; relaxation; defects; low coordination sites; segregation (multi-element); adsorbed layers; reaction profiles
Ceramics (including high T superconductors)	Grain structure; phase separation; intergranular regions; fracture faces; pores; amorphous regions; triple points; reaction products; elemental distributions; leaching profiles
Minerals (including soils, oxides, salts)	Grain structure; phase separation; altered surface layers (weathered, deposited, oxidised, reduced); faceting; reconstruction; defect sites (low coordination); intergrowth structures; leaching profiles; adsorbed layers (minerals processing)
Glasses	Hydrolysed layer; crystalline regions; inhomogeneities; elemental segregation; reaction products (distribution); mould materials transfer
Polymers	Altered surface layers (oxidised, reacted); excluded layers (lubricants, catalysts, unreacted monomer); phase separation (elastomers, monomers); surface segregation of reactive groups
Composites and natural materials (including wood, paper, paints, cement, fibreglass, tissue, blood, bone)	Distribution of materials; Modified surface layers (bonding compatibility); interphase regions; elemental distributions; reaction profiles; leaching profiles

Table 2.

Depth	Examples of the definitive role
Outer monolayer ~0.1 nm	Heterogeneous catalysis, surface tension (contact angle) control, selective adsorption, electrochemical systems, biological systems, sensors
Thin film ~0.1–100 nm	Emulsions, tribological (friction) control, anti-reflection coatings, lipid membranes, Langmuir–Blodgett films, interference filters, release agents
Near surface 0.1–10 μm	Semiconductor devices, surface hardening, polymer biodegradation, controlled release membranes, osmosis devices, photographic film, optical recording media, aerosols
Thick layers >10 μm	Anti-corrosion coatings (anodizing, electroplates, paints), phosphors, adhesives, electrowinning, magnetic recording media, surface cladding

and structure of the thin film or coating. Hence, we will need techniques that are highly surface-specific, analysing depths less than a nanometer, together with techniques capable of either collecting information from depths on the micrometer scale or profiling through a surface layer at different depths.

These techniques must also provide a description of the composition of the surface layers including an assessment of the chemistry of the thin film or coating. Impurity detection down to ppm, or even ppb, may be required. The chemistry of oxidation or environmental reaction products may need to be defined. Validation of the nature of the interfacial bonding between the layer and the substrate, often pre-reacted or plasma-treated, is sought in relation to adhesion and disbonding characteristics. All of this type of information requires chemical characterisation averaged over the surface and, in some cases, within very small regions where compositional differences may be found.

Information of importance in characterisation and QA also includes assessment of the topography, surface roughness and defect sites down to the atomic scale. Ceramic films require definition of coverage, continuity, grain size and through-film variation of these characteristics together with the phase structure of the film or coating. For layers deposited by adsorption or LB processes, the structure of monolayer, bilayer or multilayer deposition is often of critical importance in subsequent application in electronic devices, membranes etc. Hence, techniques that are able to image from hundreds of micrometers down to atoms are required ideally with ability to discriminate between different chemical species, domain structures etc.

3. SURFACE ANALYTICAL TECHNIQUES

The surface analytical techniques suitable for analysing the composition and structure of these surface layers have been extensively reviewed in several previous publications, e.g. references 1–5. The application of surface analysis methods in materials science has been surveyed in a recent compilation where a summary of techniques, together with processes and applications in materials technology, catalysis, electronic devices, oxidised metal surfaces, thin film analysis and adsorbed layers is presented.[1] In my previous contribution to the first Australian International Conference on Surface Engineering, a summary of six techniques providing analysis of the surface and near-surface regions was given. The surface-specific techniques of X-ray photoelectron spectroscopy (XPS or ESCA), scanning Auger microscopy/spectroscopy (SAM, AES) and secondary ion mass spectroscopy (SIMS) together with the near-surface techniques of analytical scanning and transmission microscopy and Fourier transform infrared spectroscopy (FTIR) were discussed with applications in study of the chemical and physical mechanisms causing degradation of the surfaces of oxides, ceramics, glasses and polymers.[2] Tables 3 and 4, reproduced from that article, summarise the characteristics and information available from these six techniques.

Table 3. Surface techniques – characteristics of XPS, Auger spectroscopy, scanning Auger microscopy and SIMS

X-ray photoelectron spectroscopy (XPS) provides
Quantitative surface compositions (To ~±10%) for all elements from He to U down to ~0.1 atomic % at depths of ~0.4–5 nm
Chemical states, i.e. bonding, valency, of elements in the surface, including proportions of the same element in different forms, e.g. oxide/metal; sulphide/sulphate, etc.
Quantitative depth profiles for all elements from angular resolved (non destructive) spectra and from ion etching
Relatively large analysis area (25 μm–10 mm).
Auger spectroscopy and scanning Auger microscopy provide
Quantitative surface compositions (to ~±20%) for elements He to U at depths ~0.4–3 nm
Physical imaging of surfaces
Spatial distribution and chemical mapping of elements to ~5 nm–0.5 μm resolution
Spot analysis of regions and crystallites
Compositional profiling
Limited chemical state differentiation
Secondary ion mass spectrometry (SIMS) provides:
High sensitivity detection of surface elements and trace impurities to ppm levels for all elements from hydrogen to uranium over large dynamic range $(1-10)^{7}$
Monolayer depths of analysis
Rapid depth profiling for thin films, surface coatings, surface layers
Identification of grain boundary and fracture face segregants using static SIMS profiling
'Fingerprint' identification of polymers and their surface reactions using static SIMS

The reason for the extreme surface sensitivity of XPS, SAM and SIMS, associated with the mean free path of the emitted electron or ion beam in the solid, was described together with the experimental basis for each of these techniques.[1,2] It is noted that XPS is particularly useful for defining the chemical state of elements in the surface layers (e.g. valency, oxidation state, compound type etc.). However, the analysis area for XPS is relatively large in many instruments (i.e. 2–10 mm) although the more modern instruments can achieve small spot analysis in the 5–25 μm range depending on the particular instrument. SAM, whilst generally only providing very limited differentiation of chemical states, is useful for rapid small spot analysis (i.e. 5–25 nm) together with mapping of the lateral distribution of elemental species across the surface. There are limitations in the Auger technique with the analysis of insulating materials and degradation of some materials (e.g. some polymers) under the electron beam as in electron microscopy. SIMS is a highly sensitive technique able to detect ppm and, in some cases, ppb levels of impurities, including hydrogen. Statis SIMS is now extensively used for identification of polymer materials, the study of surface reactions on polymers[4,5] and mapping the lateral distribution of adsorbed species, e.g. collector molecules on mineral surfaces.[6] All three techniques can be used in depth profiling mode from which the compositional changes of a thin film or coating with depth through the layer can be identified. With qualification due to damage caused by the ion beam used for sputtering, differences in the chemical

Table 4. Near surface techniques – characteristics of analytical SEM, TEM and FTIR

Analytical scanning electron microscopy (SEM/EDS) provides
Secondary electron images of surface features to (normally) ~5 nm–0.1 μm resolution
Backscattered electron images of phase differentiation, precipitates, reaction regions etc. based on average atomic number contrast
Topographic images of pits, protrusions, reacted regions etc.
Scanning maps of elemental distributions in near-surface region normally for elements above B
Quantitative analysis, with standards, for comparison with assays, XPS etc
Analytical transmission electron microscopy (ATEM) provides
High resolution (i.e. <0.2 nm) imaging of surface features, e.g. defects, facetting, steps, ledges, pits
Analysis of small particles (i.e. <0.5 μm)
Intergranular boundaries at films, specific attack
Preferential dissolution or surface reaction, e.g. specific facetting, phase preference, boundaries etc.
Precipitation and recrystallisation *in situ*
Phase identification (compositional; and structural) in grains by EDS and selected area electron diffraction
Fourier transform infrared spectroscopy (FTIR) provides
Molecular structural information on adsorbed species
Evidence of reaction in surface layers and changes in chemical species
In situ studies of adsorption and reaction on solid surfaces in solution, air or gaseous atmospheres
Rates and extents of hydrolysis in surface layers

states of particular elements at different depths through the layer can also be obtained in XPS depth profiling.

These three techniques taken together can provide most of the information required on composition and surface chemistry of thin films and coatings at the level of tens of nanometers or larger. There is still a requirement for this type of chemical information at the atomic and molecular level. Surface reactions can be followed with these techniques and the chemical processes involved at the layer/substrate interface (e.g. reaction, bonding, interdiffusion etc.) are available from studies of sequential deposition of very thin films or layers, ion beam depth profiling and angle resolved XPS (non-destructive).

The analysis of the near-surface region for both chemical and structural information can be largely achieved with a combination of analytical electron microscopy and FTIR. The new generation of high resolution field emission scanning electron microscopes with energy dispersive spectroscopy (EDS) analysis can resolve features down to ~2 nm although the EDS analysis area is generally larger (i.e. 50–100 nm). Structural information is obtained using TEM and phase contrast microscopy at the thinner near-surface regions of small crystallites or thinned sections in transmission mode. Chemical state information is, however, difficult to obtain from either EDS or electron energy loss spectroscopy (EELS) so that compositional elemental percentages are normally the limit of information available from different depths (using altered electron beam voltages). The complementary near-surface technique of FTIR, in diffuse reflectance, specular reflectance or attenuated total reflectance modes, can be used *in situ* to obtain molecular structural information. With sufficient surface area, this technique can potentially provide much of the chemical state information required on the properties of the film beneath the surface layers (i.e. 0.5–2.5 μm). It is particularly valuable in allowing observations of reactions in progress under solution or in solid-gas systems, a difficulty with the *ex situ* vacuum techniques of XPS, SAM, SIMS and electron microscopy.

To add to these six techniques, two relatively new advanced surface analytical techniques have become widely available in the last few years. They comprise scanning tunnelling microscopy/spectroscopy (STM) and atomic force microscopy (AFM). The main characteristics and types of information from these techniques are presented in summary form in Table 5. It is useful to briefly review the experimental

Table 5.

Scanning tunnelling microscopy/spectroscopy (STM) provides
3-dimensional atomic-scale imaging of ***conductive*** surfaces in air, vacuum or liquid
Topographic images of steps, ledges, vacancies (atomic) and grains, surface roughness (nm–μm scale)
Fourier-transformed diffraction structure of surface phases from atomic positions in successive layers of film or coating
Direct observation of surface reactions *in situ*, e.g. oxidation, corrosion, hydration, adsorption/desorption, precipitation etc.
Lateral distribution of adsorbed species, reaction products, i.e. patches, continuous
In spectroscopic mode, conductivity of the tunnel gap, surface electronic structure and element identification at atomic scale
Electro-chemical and electric field mapping of surface reactions and device tracking
Atomic force microscopy (AFM) provides
3-dimensional atomic scale topographic imaging and structural analysis of ***non-conductive*** surfaces in air, vacuum or liquid (as for STM)
Surface roughness imaging and statistical analysis (as for STM)
Direct observation of surface reactions and lateral distributions (as for STM)
Force–distance measurements for *in situ* particle–particle interactions
Lateral force (friction) imaging of surfaces
Non-contact (soft) imaging of bubbles, cells, LB films etc.
Magnetic domain structuring and distributions

basis of these techniques[7] in order to explain the origin of the information they provide. Scanning tunnelling microscopy uses a sharpened metal tip with a single atom protruding furthest, brought close enough to a conductive surface that electrons may quantum mechanically tunnel across the narrow gap (0.5–1.5 nm). A bias potential (1 mV–3 V) is supplied to initiate the tunnelling. The tunnel current is a strong exponential function of the tunnel gap and also depends on the bias and the inverse decay length of the tip and surface wave functions which, in turn, depend on the work functions of the atoms in these surfaces. The strong exponential dependence of the tunnelling current allows the tip height above the surface to be kept constant in feedback mode using piezo-electric crystals. Alternatively, the tip can be rastered in constant height mode and differences in the tunnelling current monitored. Imaging is non-destructive since the tip is not normally in contact with the surface during the scans. The lateral dimensions of the rastered image area can be widely varied to image from individual atoms through to regions up to hundreds of micrometers. Spatial variation of the surface conductivity and electronic structure can be used, in some cases, to distinguish between different atoms in the surface layer. Digitally-stored images of the atomic positions can also be transformed into diffraction structures allowing identification of surface phases, another very important feature of this technique applied literally to the top layer of atoms. With AFM, STM can provide ultimate information on topography, defects and surface roughness together with direct observation of surface reactions.

The AFM is essentially the complement of the STM for application to non-conductive surfaces but it can also be used for conductive surfaces providing both similar and complementary information. In this technique, an optical level system is used to measure the displacement of a ceramic tip, caused by force–distance changes, tracing across a surface. The deflection of the tip can be related, through a spring constant, to force, potential and magnetic field images from atomic dimensions to hundreds of micrometers. The technique is also particularly useful for mapping friction-related lateral force differences across surfaces and magnetic domain structures and distributions. Comparison between STM and AFM images from the same surface has demonstrated the complementarity of the two techniques.[7,8] For work on biological applications and soft LB and adsorbed layer films, this technique has significant advantages and, in some cases, may be the only technique likely to give information relating to the unaltered original structure of the film.[8]

Rather than belabour the general features of these two techniques, it would probably be more useful to the reader to give some specific applications of STM and AFM in which some of the information listed in Table 5 is derived from particular surface engineered layers.

4. CASE STUDIES

In this section, some examples of the applications of XPS, SAM, AFM and STM will be described. References to the use of the other techniques can be found in the previous article[2] but some additional comments will be provided in this section for particular techniques to which these systems have been applied.

4.1 Studies of a Ni/TiN/Mildsteel Coating System

A mild steel substrate coated with titanium nitride using PVD was overcoated with a deposited layer of electroless nickel. A taper section was then cut through the two coating layers into the steel substrate and prepared in metallographic cross section for examination in the SAM. The secondary electron image obtained with the SAM (Perkin Elmer PHI 600 system) is shown in Fig. 1 with the magnification of 100 μm indicated by the white bar in the micrograph. The steel substrate is the section in the top third of the figure, the taper cut through the titanium nitride is in the centre third with the electroless nickel layer shown in the lower third of the figure. The uneven boundary between the titanium nitride and nickel layers is clearly indicated in the secondary electron image. Figure 1(a) shows the distribution of oxygen across this section. This map is derived from the O KLL Auger signal gated to the detector so that only signals from this element are recorded. The titanium nitride layer is obviously highly oxidised compared with the steel and electroless nickel layers. The structure of the interface is less obvious in the oxygen map but some features are retained, suggesting that there is no significant interdiffusion between the titanium nitride and nickel layers. This result is confirmed in Fig. 1(b) where the Ni LMM signal is used to obtain a nickel distribution in the same layers. The titanium nitride/nickel boundary is closely defined and shows features similar to that in the secondary image of Fig. 1. It is interesting to note, however, that there is a small inclusion of nickel near the steel/titanium nitride interface in the top left-hand corner of Fig. 1(b). This appears to have been embedded in the steel surface before coating with the titanium nitride. The presence of this nickel inclusion is confirmed by the Fe LMM map in Fig. 1(c) where the position of the nickel inclusion is seen as a region not containing iron at the same point in the micrograph. There are some indications in the Fe map that iron may have diffused in small concentrations into the titanium nitride and, possibly, electroless nickel layers since there appears to be a concentration gradient moving towards the surface from the steel substrate interface but this would require confirmation (see below). The Ti LMM map is shown in Fig. 1(d) and again there are some indications of interdiffusion of titanium into the steel substrate and possibly into the electroless nickel layer since the interface for this element is considerably less well defined than those seen for oxygen and nickel. This type of elemental mapping can be carried out at different magnifications down to sub-micron levels of detail.

A line scan from the top to bottom of Fig. 1 across the centre of the micrograph is illustrated for Fe, O, Ti, Ni in Fig. 2. A line scan for N was not carried out since this signal is partially overlapped with the Ti signal in Auger spectra. The Ti signal therefore contains some contribution from N in this figure. It is interesting to note that the surface layers of the steel show a significant concentration of oxygen and this result is in accord with Fig. 1(a) where there is a marked difference in oxygen content between the steel substrate and the electroless nickel layers. The line scan confirms that the highest level of oxidation is found in the titanium nitride layer. The electroless nickel is the least oxidised layer. The line scan does not show the presence of significant concentrations of iron in either the titanium nitride or the nickel layer suggesting that the appearance of this concentration gradient in Fig. 1(c) is an artefact. This was confirmed by separate point analyses taken within each of the three layers. No detectable iron was found in either of the two overlayers. It is therefore possible to confirm or deny apparent observations of interdiffusion by separate small spot analyses at different points across a coating or thin film as in this case. The

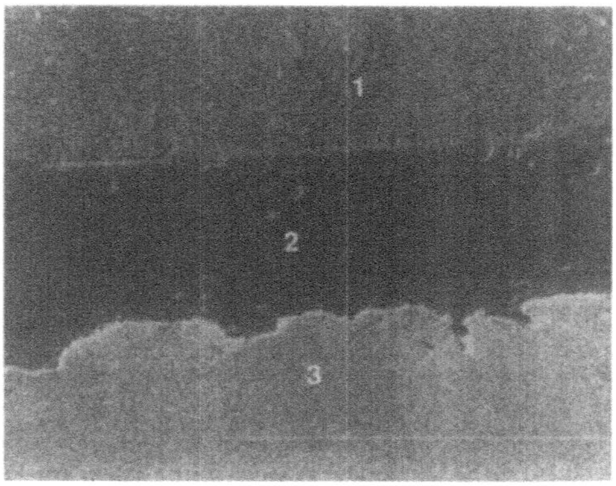

Fig. 1. Secondary electron image coatings of electroless nickel (Region 3) and TiN (Region 2) on mild steel substrate (Region 1).

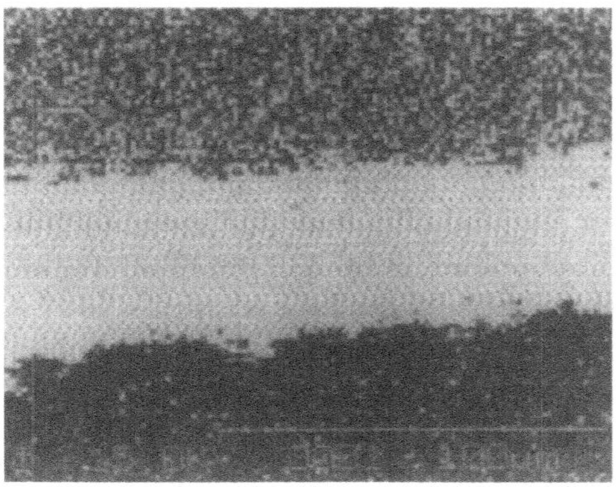

Fig. 1a. Oxygen (KLL) scanning Auger map (SAM) of Fig. 1.

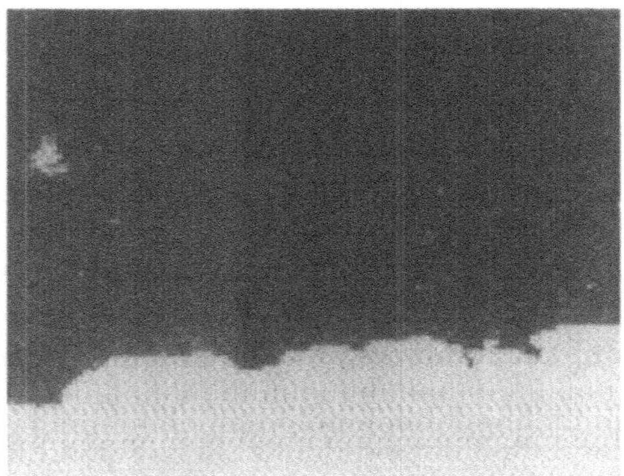

Fig. 1b. Nickel (LMM) SAM of Fig. 1.

Fig. 1c. Iron (LMM) SAM of Fig. 1.

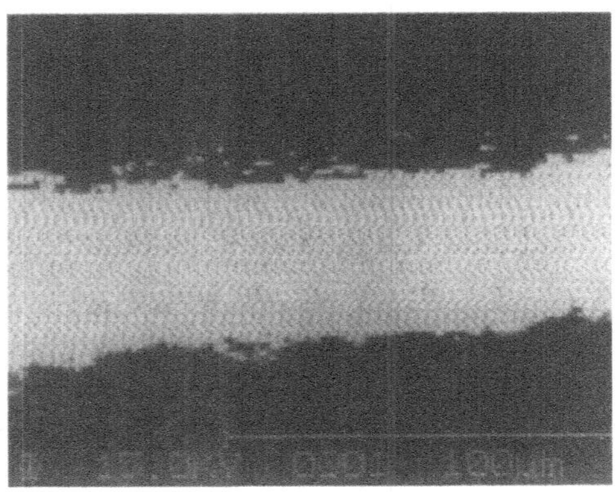

Fig. 1d. Titanium (LMM) SAM of Fig. 1.

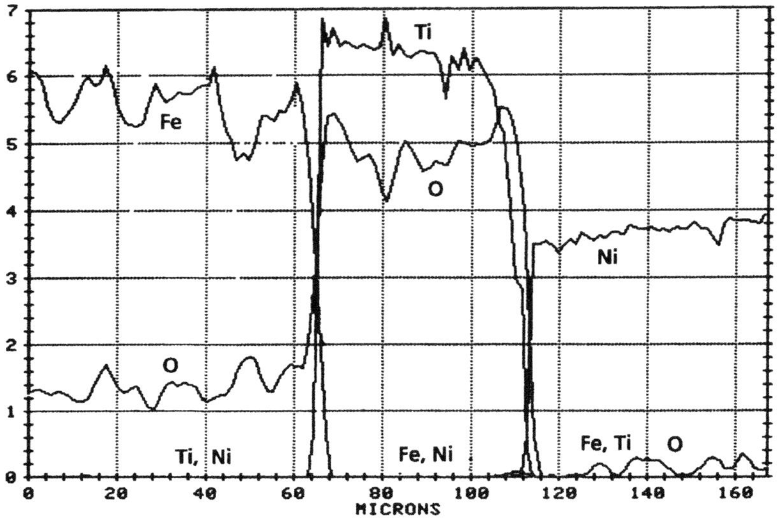

Fig. 2. Line scans for Fe, O, Ti and Ni down Figure 1 (centre, top to base).

separate point analyses within the titanium nitride layer did, however, reveal the presence of small concentrations of impurities of aluminium and sodium apparently derived from sputtering processes during the PVD coating process.

At the other extreme in characterisation and QA, Fig. 3 illustrates a poor example of a similar system of steel/titanium nitride/electroless nickel coating in secondary electron imaging mode. It is immediately apparent that there are a number of major defects comprising voids, impurities, particulates, reaction layers at interfaces and deposited particles in the metallographic section. In principle, it is possible to analyse the individual particles as well as the separate layers as in the previous example. The reason for the failure of the coating system in this case was, however, readily determined by point analyses in each of the three layers where the presence of sulphide was detected in all three layers as illustrated in Fig. 4 for the nickel and titanium nitride layers. The presence of this impurity in the gas phase during PVD processing has led to the poor structure of the coating layers.

In summary, Auger spectroscopy can be used in these modes to give composition, stoichiometry and impurity measurement together with depth profile and interfacial bonding information. The practical limit of detail is, however, in the 50–100 nm range.

4.2 XPS of Titanium Nitride PVD-Coated Mild Steel

A second example in which the compositional information derived from one surface analytical technique is convoluted with the requirement for structural information derived from another technique can be seen in XPS studies of depth information through a titanium nitride PVD coated mild steel sample. The Auger studies in the previous section, together with a considerable literature (e.g. ref. 9 and references therein), has shown that the titanium nitride (e.g. ref. 10) surface is extensively

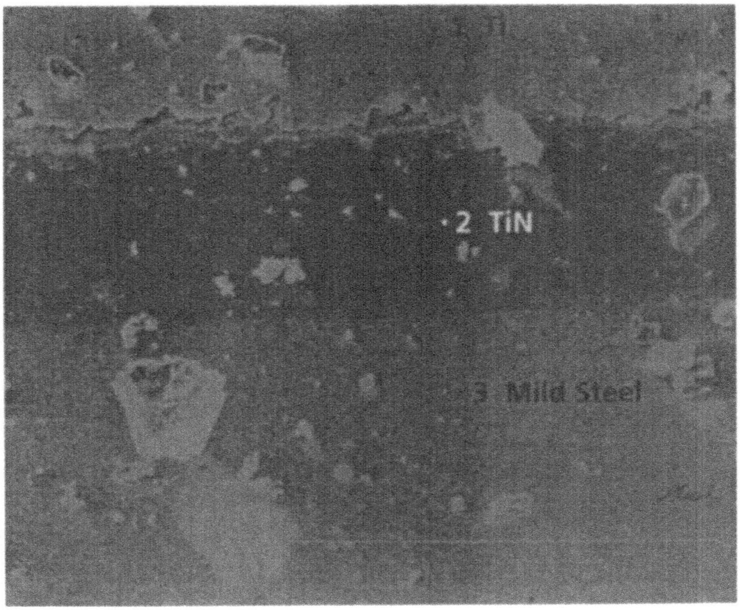

Fig. 3. Secondary electron image of poor Ni/TiN/mild steel coated system.

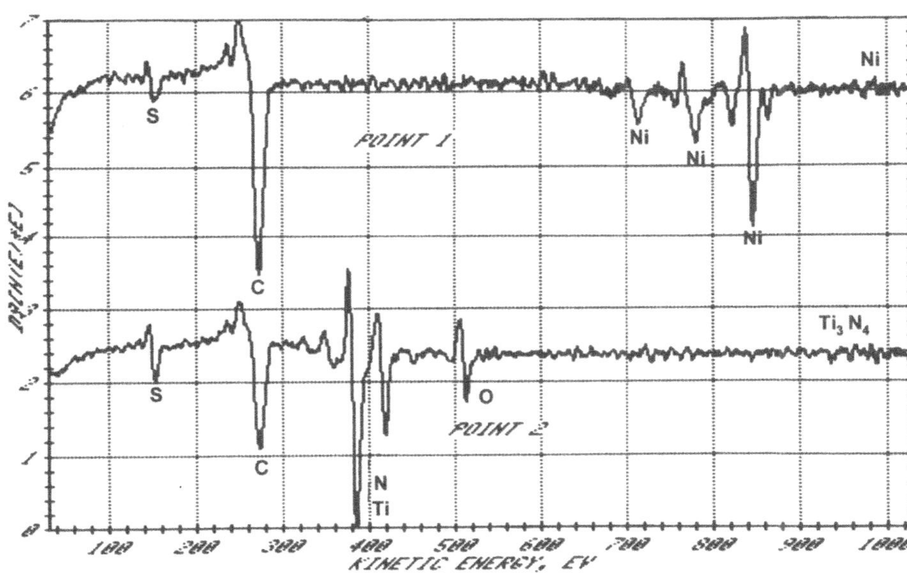

Fig. 4. Auger survey spectra from nickel and titanium nitride layers in Fig. 3 showing the presence of sulphide impurity.

oxidised by exposure to air. The XPS technique has the significant advantage of being able to separately present the signal both from titanium and nitride. Figure 5 shows the XPS spectra from the N1s photoemission as a function of argon ion etch

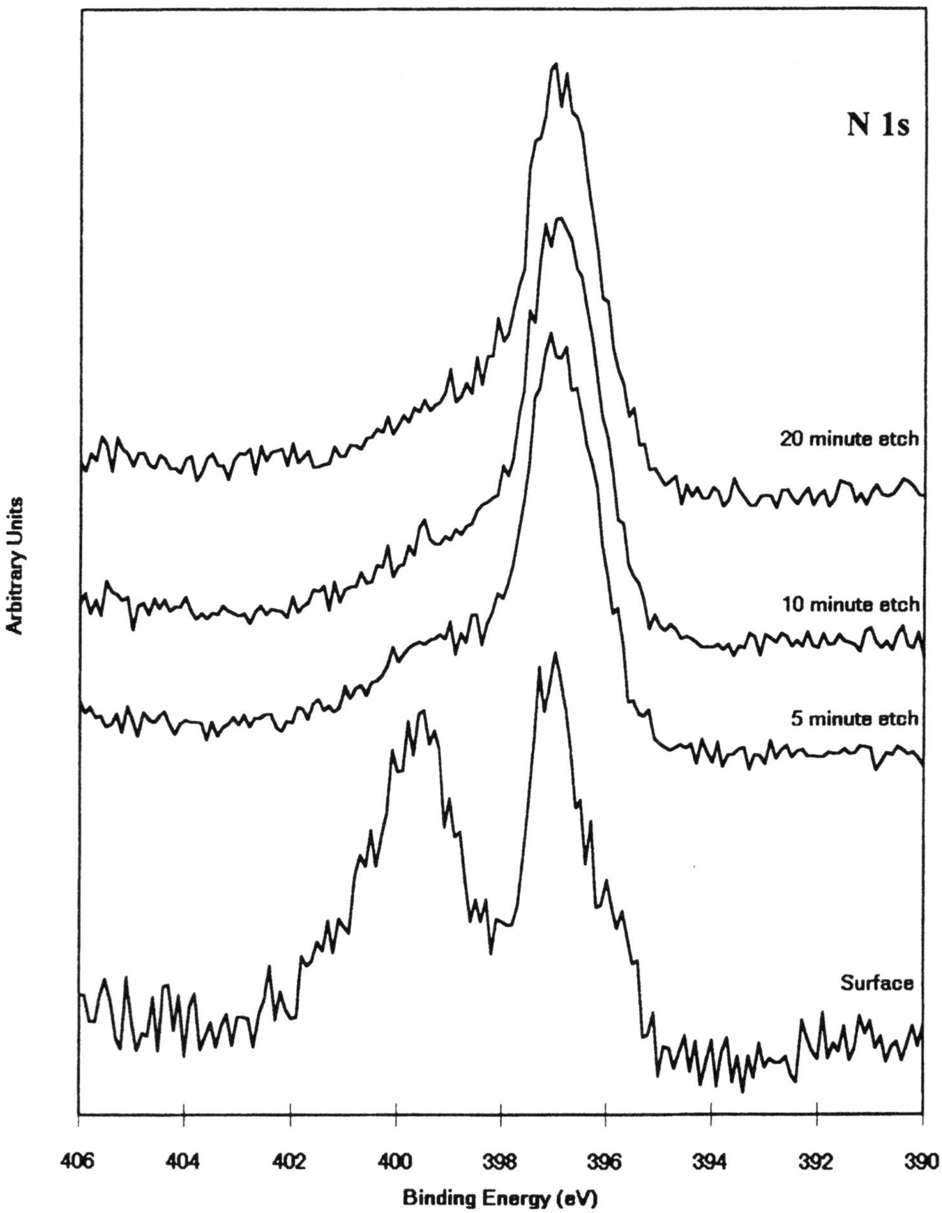

Fig. 5. XPS spectra of N1s region from a TiN coating on mild steel subjected to argon ion etching for 0, 5, 10, 20 min.

time. The ion etch rate corresponds to the removal of approximately 2.5 nm in five minutes in a 10 × 10 mm rastered region. The initial surface, before ion etching, shows signals from both the nitride near 397 eV and an oxidised nitrogen species near 400 eV attributable to either adsorbed nitrogen or $-NH_2$ species. This species is rapidly removed from the surface layers together with some of the oxygen in the elemental composition. The Ti2p spectral region also shows the presence of titanium in two chemical states as the nitride and oxide. The $Ti2p_{3/2}$ peak can be curve fitted into components corresponding to these two species and the separate intensities of these components monitored as a function of ion etch depth. A depth profile of this type is illustrated in Fig. 6. The carbon signal, as both hydrocarbon and graphitic

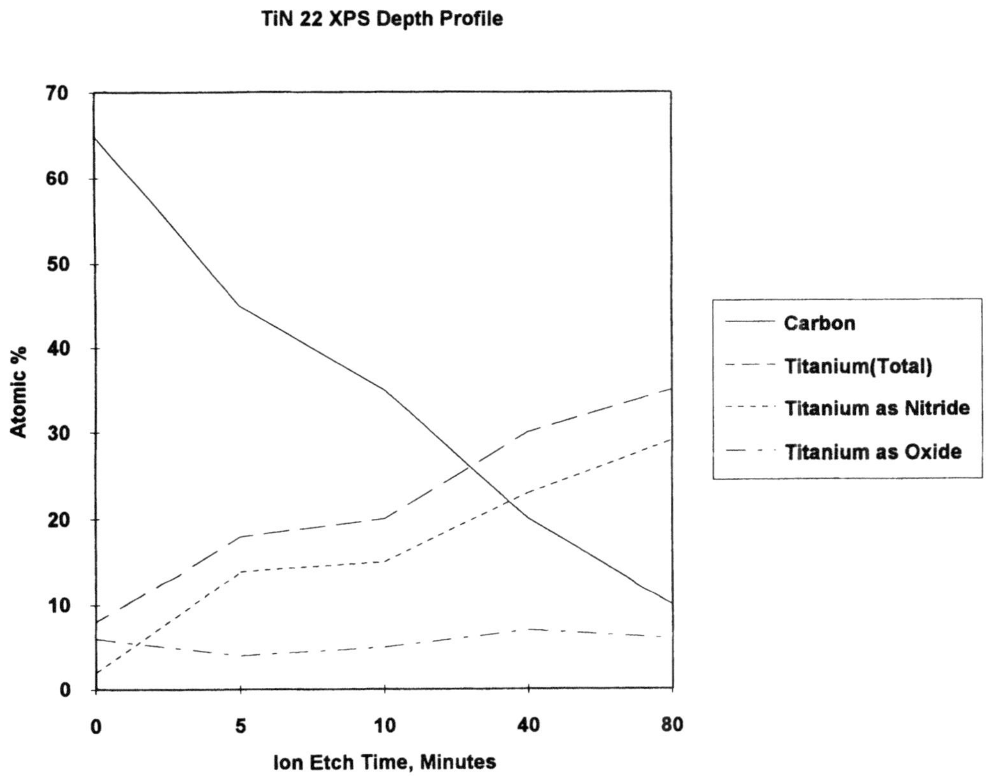

Fig. 6. Depth profile for titanium Ti2p signal oxide and nitride through TiN layer on mild steel. Note continuing presence of oxide component through coating.

contamination, only slowly declines over an 80 min ion etch period. The atomic percentage of titanium increases steadily over this period in parallel with the declining carbon concentration. The component of the titanium signal as nitride increases similarly to the total titanium but it is significant that the oxide component of the titanium signal is not removed by this ion etch over the total period. In fact, the oxide component becomes essentially constant whilst the ion beam is sputtering through the titanium nitride layer.

The reason for this, at first sight, surprising result is learnt when the structure of the titanium nitride layer is studied using metallographic and electron microscopic techniques. Examples of this structure have been given in another Chapter in this Volume and in literature.[11] In essence, a columnar structure illustrated schematically in Fig. 7, results from the PVD process. This structure retains internal porosity in

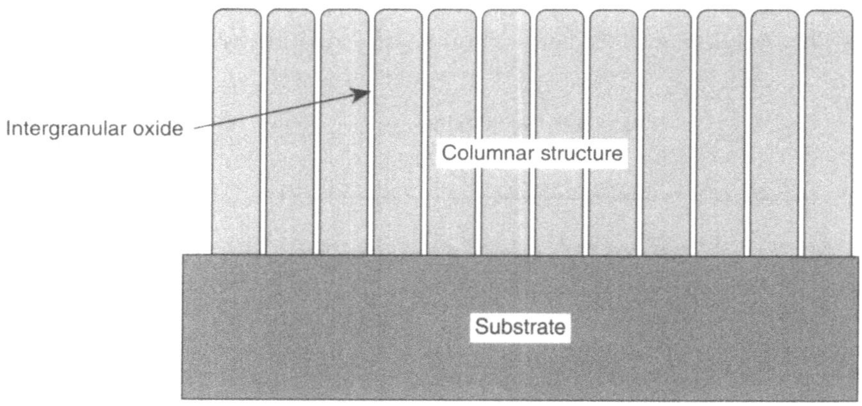

Fig. 7. Schematic illustration of columnar growth of TiN layer showing intergranular regions in which surface oxidation occurs through the film.

roughly vertical, elongated annuli. The XPS results show clearly that intergranular oxidation can occur in this porous film throughout the titanium nitride layer. The continuing presence of the oxide peak in the $Ti2p_{3/2}$ signal can only reasonably be attributed to this intergranular oxidation of the columnar structure. In general, argon ion beam etching may lead to reduction of TiO_2[12] but would not be expected to produce oxidised species or, particularly, a plateau value for the atomic concentration of the oxidised species.

In depth profiling using ion beams, knock-on vertical interlayer mixing can give rise to complications in the definition of the interfacial regions between two layers. If the analysis is being carried out over a relatively large area, as is usual in XPS depth profiling, the structure of the interfacial region can also contribute broadening of the interfacial profile. This is clearly illustrated in the convolutions seen at the titanium nitride/electroless nickel interface in Fig. 1 if an area-averaged region of analysis or rastered region is used. It is therefore desirable to use small spot Auger analysis for more precise definition of the interface when this is possible. For titanium nitride films, this analysis is made difficult by the direct overlap of the N KLL and Ti $L_3M_{23}M_{25}$ Auger peaks near 387 eV kinetic energy. These peaks are sufficiently overlapped to make any definition of separate signals impossible without further processing. Target factor analysis[13,14] can, however, separate the concentrations of nitrogen and titanium if the line shape of the titanium nitride spectrum does not change throughout the layer. It has been shown, for instance, that a depth profile of titanium nitride on titanium metal can be calculated using this technique.[14] Quantitative analysis of nitrogen and titanium content for differences in stoichiometry, caused by different processing conditions, can now be obtained from standard samples. The construction of the data set, followed by principal component analysis will allow the determination of depth profiles for titanium and nitrogen atomic concentrations. Comparisons with Rutherford backscattering analyses have validated the factor analysis approach to Auger spectra.

XPS can, particularly with small spot and imaging facilities, give a large proportion of the information required on the composition and chemical states of elements in coatings and thin films together with characterisation of oxidation, reaction and interfacial bonding.

4.3 AFM of Silane-Modified Oxidised Silicon Surfaces

Silanes are used in commercial operation as a pre-cladding agent for silicon chips before the application of a moisture protective epoxy coating. The structure and chemistry of the silane layer are both of critical importance in the protection of the silicon wafer from corrosion in electronic devices. It is essential that the silane adsorbs on the silicon surface as a complete monolayer with closely packed molecules, i.e. high adsorption density. The monolayer should be stable and irreversibly adsorbed in order that the epoxy coating is fully protective of the silicon surface.

Octadecyltrichlorosilane (OTS) is a commonly used reagent in the formation of these modified surface layers. The reaction of OTS with the oxidised silicon surface proceeds in two steps. The first is the hydrolysis of the Si–Cl bonds to Si–OH species. Some of the Si–OH species are then proposed to react with similar species on the surface of the oxidised silicon to eliminate water forming a Si–O–Si covalent link to the silicon surface. However, it is recognised that most of the reactive OH groups on the silane bond to each other to form a laterally polymerised surface network.[15] The polymerised monolayer, with some bonds to the silicon surface, thus provides the tightly packed monolayer required for coating with the epoxy outer layer.

The OTS adsorption was carried out on a 1 cm^2 section of Si wafer allowed to oxidise in air to an equilibrated layer. Solution concentrations of OTS in hexadecane were allowed to contact the wafer for 30 min after which the sample was rinsed 10 times in carbon tetrachloride, 10 times in ethanol and a final 10 times in carbon tetrachloride. The complete monolayer was then cured overnight at 150°C to maximise covalent linkage.

The characterisation and QA required to monitor the successful formation of the layer can be illustrated using atomic force microscopy. The topography of a complete monolayer of OTS on the oxidised silicon surface, formed from 5×10^{-3} M OTS solution, is shown in Fig. 8. It is noted that the roughness of the surface after monolayer formation, illustrated by the contrast changes in the bar on

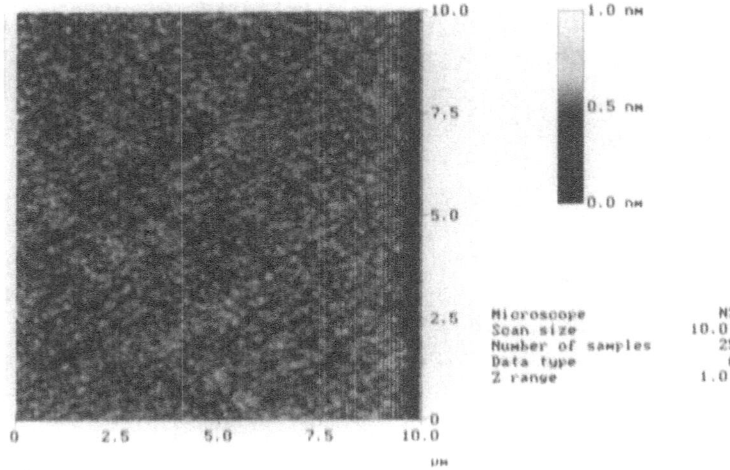

Fig. 8. AFM of complete monolayer of OTS (5×10^{-3} M) on oxidised silicon wafer. Note height contrast <1 nm (bar) over $10 \times 10\,\mu m$ area.

the right-hand side of the figure, is <1 nm. The imaged region of $10 \times 10\,\mu m$ can be varied using the AFM up to 125 µm or down to levels of atomic detail.

For comparison, if the monolayer is incomplete or the adsorption process has been inhibited by impurities or other solution effects, the AFM can give this information in direct form. The topography of the partial monolayer formed from 5×10^{-4} M OTS is shown in Figs 9 and 10. In Fig. 9, the silicon surface was removed from solution, dried in air and imaged in air. The formation of separate roughly spherical aggregates of silane molecules, with diameters of the order of 300–500 nm, can clearly be seen on this surface. The height of the separate aggregates is roughly 2 nm,

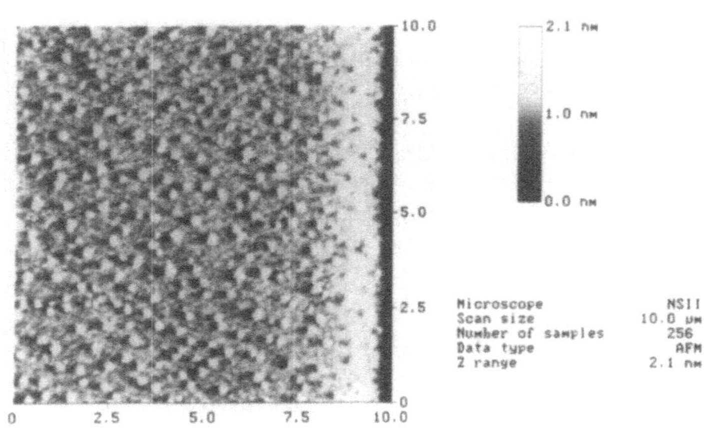

Fig. 9. AFM of partial monolayer (5×10^{-4} M) of OTS on oxidised silicon, dried in air.

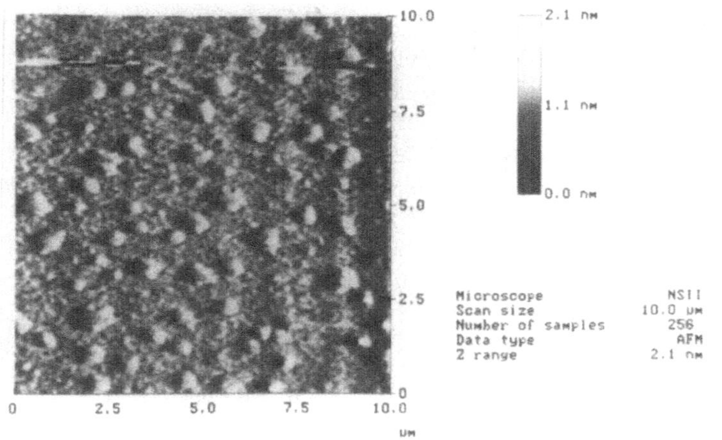

Fig. 10. AFM of same partial monolayer (5×10^{-4} M) imaged under water. Note larger (>2 nm) height variation.

i.e. a considerably rougher surface than that found after formation of the complete monolayer.

It is also possible to image the partial monolayer in aqueous solution as in Fig. 10. The presence of the water has increased the aggregate or island diameters to 0.8–1.0 µm. The islands have apparently rearranged into a lower energy state in the presence of the solution, inferring that most of the silane molecules are not covalently attached to the surface but are able to aggregate into the spheroids seen in the two images. This is confirmed in Fig. 11 where a partial monolayer formed in a lower concentration of 1×10^{-4} M OTS shows network or dendritic growth

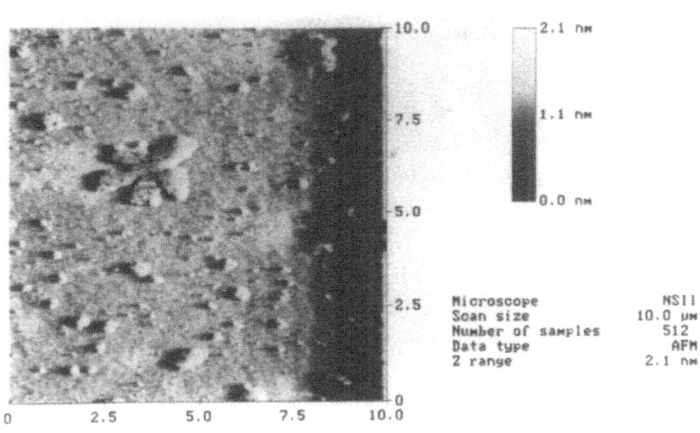

Fig. 11. AFM of unstable OTS partial monolayer (1×10^{-4} M) in water forming networks and aggregates.

indicative of formation of a crystal-like structure. The silane molecules have a greater affinity for each other than the oxidised silica surface. The formation of the full monolayer is therefore contingent on close packing and lateral polymerisation of the silane molecules. Further studies of the interaction forces between the coated silicon surfaces have also been undertaken using the AFM. A 5–10 μm glass sphere was attached to the end of the AFM cantilever using an epoxy resin. The glass sphere was then moved systematically towards the surface of the silane-coated oxidised silicon surface in order to obtain force–distance information on the interaction forces. This can be carried out in aqueous solutions in order to provide information on hydrophobicity, hydration and surface charging effects including specific adsorption of ions from the solution. The results will not be described in detail here but it is noted that the OTS modified silicon surface remains charged. In NaCl solutions at fixed pH, increasing the salt concentration increases the force/radius ratio up to 10^{-3} M NaCl. Variation in force/radius ratio with pH is found as expected from a theoretical model (i.e. DLVO theory) of this surface.

A large variety of adsorbates and reagents have also been used to chemically modify the surface of silica and other oxides for incorporation as fillers and fibres in reinforced composite materials and as insulation layers in electrical and thermal devices. Alcohols, silanes, surfactants and siloxanes reacted with silica surfaces have been extensively studied in our research[16–21] using FTIR and XPS characterisation. These two methods have elucidated the mechanisms of adsorption and surface reaction together with film thickness and surface coverage. For instance, in adsorption of primary alcohols to produce strongly hydrophobic surfaces, quantitative measurement of adsorption densities using both techniques has been validated.[18,20,21] Partial oxidation of these alcohols to produce polymerised products with multiple adsorption sites apparently confers stability on the hydrophobic surface layer.[16] A comparison of the film thicknesses derived from long-chain alcohols, alkyl ammonium bromide and silanes has shown that much thinner films are formed with the silane reactants despite the alteration of contact angle (i.e. hydrophobicity).[21] The XPS results[21] strongly suggested that island or patch-wise adsorption of silanes may have occurred in this system. Assessment of the relative adsorption densities and bonding strengths of different siloxane reagents has also been studied on E-glass fibre surfaces.[17,19]

With a surface showing partial monolayer coverage or surface roughness, quantitative data defining the nature of the surface roughness is usually required in order to compare surfaces prepared under different conditions. The AFM provided by most commercial suppliers of the instruments now provides a variety of software packages for this quantification. Some illustrations of the type of data available can be provided from an analysis of partial monolayers like those in Figs 9 and 10 presented in different formats. Figure 12(a), (b), (c) give plots of similar sub-monolayer adsorption (with higher surface roughness imaged under water) presented in top view (two-dimensional), line scan and three-dimensional imaging formats. The Section Analysis data, presented in Figure 13 provides cross-sectional profiles and roughness characterisation measured along a user-defined reference line marked as

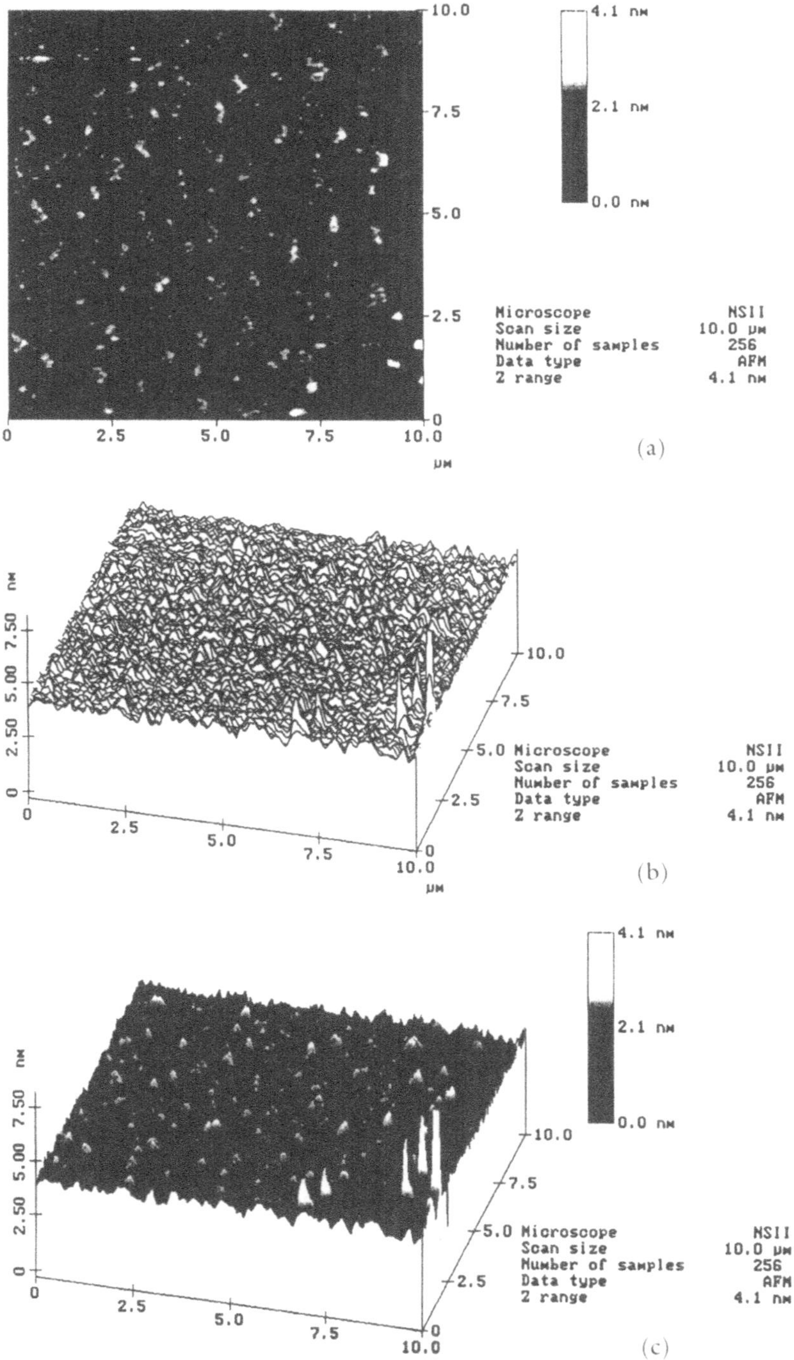

Fig. 12. Different AFM images of partial OTS monolayer (5×10^{-4} M) under water (Fig. 10) (a) Top view image. (b) Line scan image. (c) 3-dimensional image.

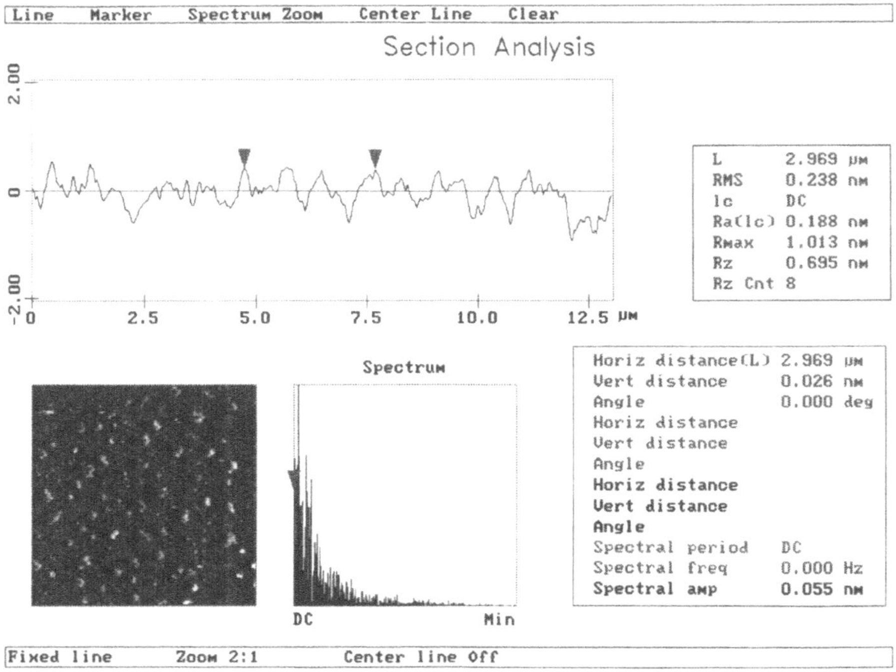

Fig. 13. Section analysis of Fig. 10 giving cross-sectional profiles and roughness characterisation parameters.

the diagonal in the figure. A sub-section (L) of the roughness curve can then be defined and the mean roughness (R_a), maximum roughness (R_{max}) and 10-point mean roughness ((R_z) derived. The RMS (standard deviation) of the Z values between the reference markers is also given.

The Bearing Analysis calculates and displays bearing ratio curves and surface–height histograms for lines or areas drawn on the image. The bearing ratio curve is the integral of the surface–height histogram and plots the total percentage of the surface above a reference plane as a function of the depth of that plane below the highest point in the image. In Fig. 14 the box area of the 10 × 10 μm region is analysed. The centre line average is the depth at which the number of points above and below the depth point are equal as defined for the entire image. The bearing depth is marked by the triangular cursors and, as the highest point in the image, serves as a reference point for the depth measurements. The bearing area is the area above the bearing depth and the bearing length is the length above the bearing depth. The bearing ratio is defined from the bearing ratio curve (i.e. the integral of the histogram curve) and gives the percentage of the surface above the reference plane defined by the cursor. The histogram area, length and percentage, are defined at the depths encountered by the cursor with the percentage representing the total number of points at this depth.

The Roughness Analysis in Fig. 15 calculates roughness parameters and surface

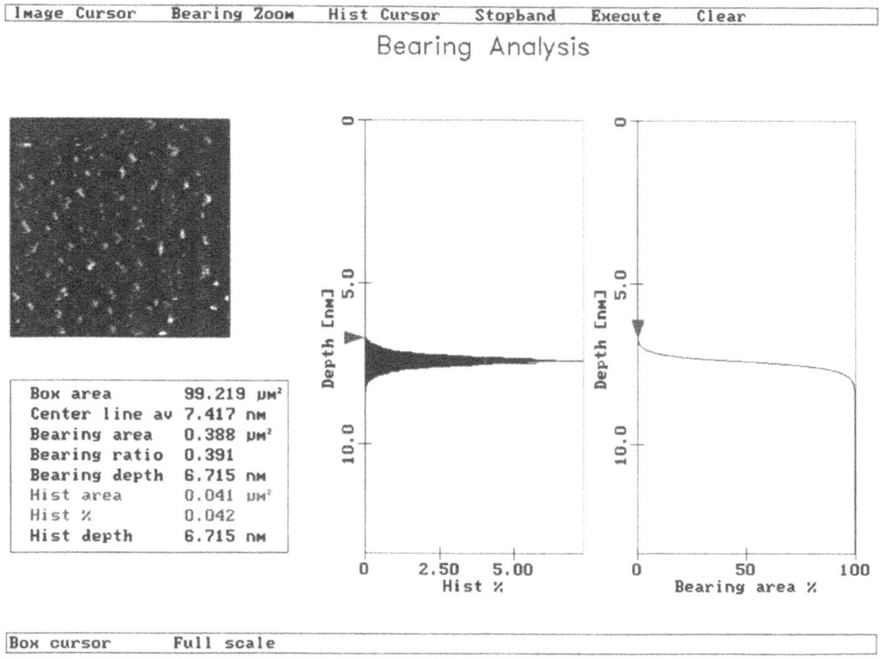

Fig. 14. Bearing analysis of Fig. 10 (see text).

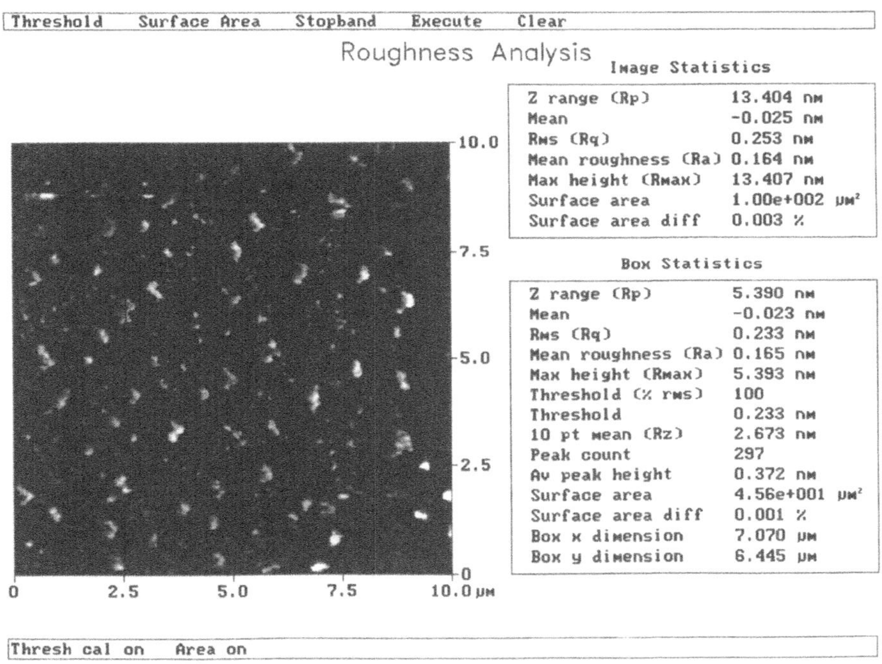

Fig. 15. Roughness analysis of Fig. 10 giving mean roughness, maximum height and surface area (see text).

areas over the user-defined area. The selected region for analysis is marked in the darker contrast in the figure. The *Z* range is the difference between the highest and lowest points within this area with the mean being the average of all the *Z* values within the enclosed area. The negative value of the mean *Z* arises because the *Z* values are measured relative to the *Z* value when the microscope is engaged. This value is not corrected for tilt in the plane of the data. The RMS (R_q) is the standard deviation of the *Z* values. The mean roughness (R_a) is the mean value of the surface relative to the centre plane and the maximum height (R_{max}) is the difference in height between the highest and lowest points relative to the mean plane. The 10 point mean (R_z) is the average difference in height between the five highest and five lowest peaks relative to the mean plane. The peak count gives the number of surface peaks higher than this threshold with the average peak height for these peaks calculated. The surface area is a three-dimensional calculation derived from the sum of all of the triangles formed by three adjacent data points in the given area.

The Grain Size Analysis (Fig. 16) defines grain boundaries based on the height

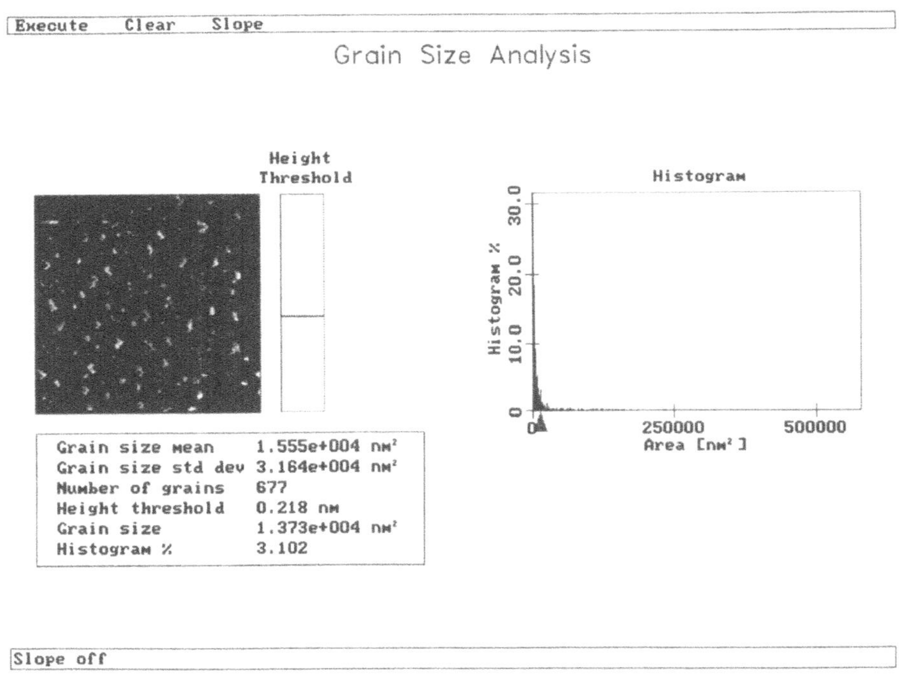

Fig. 16. Grain size analysis of Fig. 10 (see text).

and slope of the data from which a histogram of the grain size distribution can be produced. The standard deviation of the grain sizes and the total number of grains are then calculated and displayed as in the figure.

The Fractal Analysis (Fig. 17) gives a different measure of the roughness of a sample surface. In principle, the Fractal Analysis is carried out by superimposing a three-dimensional array of cubes on the three-dimensional image so that the cubes

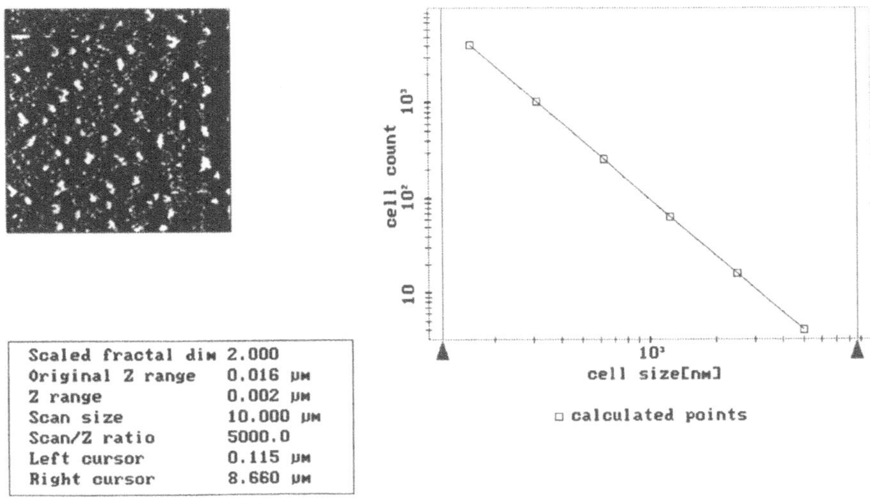

Fig. 17. Fractal analysis of Fig. 10 (see text).

completely encompass the image. The size of the cubes is then varied and the number of cubes intersected by the image is recorded for each cube size. With rougher samples, for a given cube size, more cells are intersected than smooth surfaces. The fractal dimension is defined as the slope of the line obtained by plotting the log of the cell size against the log of the cell count. Clearly, the fractal dimension increases with the roughness of the sample from a minimum of 2.0 for smooth samples to a maximum of 3.0 for an extremely rough sample. There are practical difficulties in this analysis which has led to most commercial instruments providing a 'scaled' fractal dimension in which the *Z* range parameter is changed without changing the values of the stored data in order to allow intersection of more cell boundaries. Nevertheless, with an image of the type shown in Fig. 17, the scaled fractal dimension still shows a relatively flat surface. This form of analysis is really only useful for relatively rough surfaces. Finally, with some instruments, an Autocovariance image is also calculated from the autocovariance function. This function is calculated by taking the inverse Fourier transform of the product of the FFT of the image and the complex conjugate of the FFT.

These different formats for presentation of the data will illustrate the range of information available to allow quantitative comparison between topographic detail on samples prepared by different processing conditions.

4.4 STM of Oxidation and Corrosion of Sulphide Surfaces

The oxidation and corrosion of sulphide surfaces has important consequences in fields as diverse as microelectronic devices (e.g. ZnS, CdS) and mineral separation processing.[22–26] The initiation of surface oxidation at defect sites (e.g. steps, vacancies, corners, edges) and impurity sites can be directly observed down to atomic level using scanning tunnelling microscopy). The growth of the oxidation products through to an oxidised overlayer can be followed both in air and in solution for conductive sulphide materials.

An example similar to that in previously published works[23,26] is reproduced here to

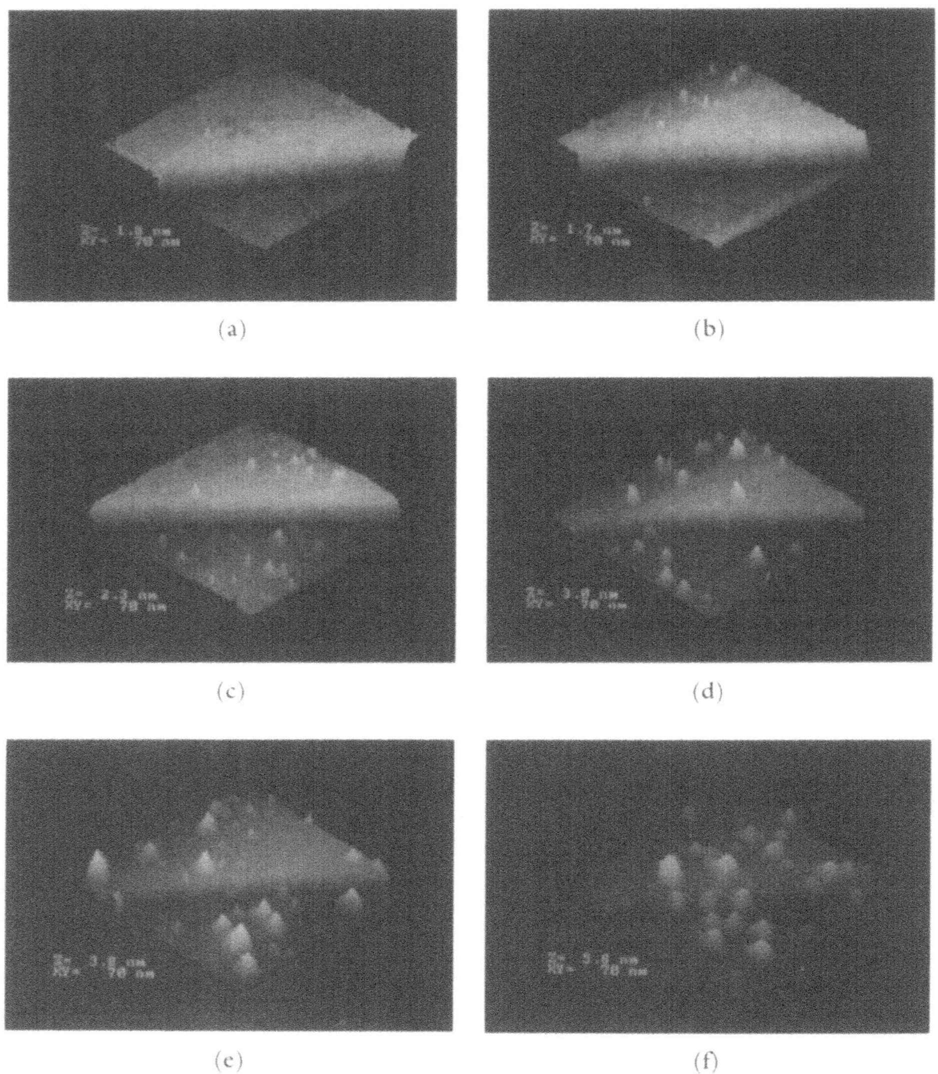

Fig. 18. Successive STM images (70 × 70 nm region) from a galena (PbS) surface oxidised in air for (a) 0 (b) 10 (c) 60 (d) 150 (f) 500 min. Note the change in Z parameter from 1.8 nm to 5.8 nm as the oxidation products grow on the surface.

illustrate the sequence of imaging available from this technique. A galena (PbS) surface, freshly cleaved and exposed to air for successive periods of time, shows the formation of separate initiation sites and oxidation regions as illustrated in Fig. 18. The separate regions grow gradually from a few atoms to areas that eventually overlap to form a porous oxidised layer. XPS identifies the initial product formed in air as lead hydroxycarbonate with later addition of lead hydroxide to the oxidation products.[23]

In solution, similar images can be obtained showing the process of oxidation and corrosion of similar form. Different pH conditions dictate whether pitting or *in situ* precipitation is the predominant mechanism of surface attack. At low pH, loss of material from the surface is found with predominantly oxide/hydroxide surface products. At high pH, similar attack to that found in air with the production of lead hydroxycarbonate and lead hydroxide can be identified with the combination of STM and XPS.[26] Full details can be found in references [23,26] but this brief sample will illustrate the ability of the STM technique to provide detailed structural information on the characterisation of mechanisms of alteration of the surface in oxidation, corrosion and surface reaction.

5. CONCLUSIONS

In this brief survey, the objective has been simply to indicate that there is now available a variety of surface analytical techniques from which most of the information on the chemistry and structure of surfaces can be obtained at levels of detail from the atomic to the millimetre scale. In particular, the more recent availability of atomic force microscopy and scanning tunnelling microscopy has greatly added to our ability to provide structural information allied to that obtained from the surface chemical techniques.

The use of these techniques in characterisation and quality assurance for surface engineering processing will undoubtedly increase over the next decade. It is likely that many of these techniques will become available in industry on an increasingly routine basis but the initial (and present) use of the techniques in diagnostic problem solving will precede the more extensive applications to quality assurance.

REFERENCES

1. 'Surface Analysis Methods in Materials Science', J. O'Connor, B. Sexton and R. St. C. Smart, eds, *Springer Series in Surface Science*, Vol. 23, Springer-Verlag, Berlin, 1992.
2. R. St. C. Smart and G. A. George: 'Surface Degradation of Oxides, Ceramics, Glasses and Polymers', *Surface Engineering: Processes and Applications*, K. N. Strafford, I. Sare and R. St. C. Smart, eds, Technomic Publishing Co., Penn., USA, 1994, 269–312.
3. W. Hirschwald: 'Selected Experimental Methods in the Characterization of Oxide Surfaces', *Surface and Near-Surface Chemistry of Oxide Materials*, J. Nowotny and L.-C. Dufour, eds, Elsevier, Amsterdam, 1988, 61–188.

4. *Methods of Surface Analysis: Techniques and Applications*, J. M. Walls ed., Cambridge University Press, UK, 1989.
5. D. R. Randell and W. Neagle: 'Surface Analysis Techniques and Applications', Royal Society of Chemistry, *Special Publication No. 84*, 1989.
6. J. S. Brinen and F. Reich: *Surf. Interf. Anal.*, 1992, **18**, 448–452.
7. *Scanning Tunnelling Microscopy and Spectroscopy: Theory, Techniques and Applications*, D. A. Bonnell, ed., VCH Publishers Inc., USA, 1993.
8. N. A. Burnham and R. J. Colton: 'Force Microscopy', *Scanning Tunnelling Microscopy and Spectroscopy: Theory, Techniques and Applications*, VCH Publishers Inc., USA, 1993, 191–251.
9. S. Hofmann: *Thin Solid Films*, 1990, **193/194**, 648–664.
10. S. Nordgren, G. Bray, Y. Claesson, M. Georgson, C.-G. Ribbing and N. Wassdahl: *Surf. Coatings Tech.*, 1990, **43/44**, 1015–1023.
11. G. Hakansson, L. Hultman, J.-E. Sundgren, J. E. Greene and W.-D. Munz: *Surf. Coatings Tech.*, 1991, **48**, 51–67.
12. S. Myhra, H. E. Bishop and J. C. Riviere: *Surf. Tech.*, 1983, **19**, 161–172.
13. D. G. Watson: 'Factor Analysis of AES and XPS Spectral Data', *Technical Bulletin 8905*, Perkin Elmer Physical Electronics Division, Minnesota, USA, 1989.
14. D. G. Watson, W. F. Stickle and A. C. Diebold: 'Quantitative Analysis of Titanium Nitride by Auger Electron Spectroscopy', *Thin Solid Films*, 1990, **193/194**, 305–311.
15. C. P. Tripp and M. L. Hair: *Langmuir*, 1991, **8**, 1120.
16. T. J. Horr, J. Ralston and R. St. C. Smart: 'FTIR Diffuse Reflectance Study of the Adsorption and Oxidation of Alcohols on Porous Silica Surfaces', 1992, **63**, 21–28.
17. D. R. Bennett, J. G. Matisons, A. K. O. Netting, R. St. C. Smart and A. G. Swincer: 'Silicones on Glass Surfaces', *Polymer International*, 1992, **27**, 147–156.
18. T. J. Horr, J. Ralston and R. St. C. Smart: 'Methods for Quantitative Diffuse Reflectance FT-IR: Adsorption Densities of Alcohols on Silica Powders', *Colloids and Surfaces*, 1992, **64**, 67–85.
19. L. G. Britcher, D. C. Kehoe, J. G. Matisons, R. St. C. Smart and A. G. Swincer: 'Silicones on Glass Surfaces, Pt. II. Coupling Agent Analogs', *Langmuir*, 1993, **9**, 1609–1613.
20. T. J. Horr, J. Ralston and R. St. C. Smart: 'The Use of Overlayer Equations in an XPS Study of the Adsorption of Long Chain Alcohol Molecules on Silica Surfaces', *Colloids and Surfaces* A, 1994, **92**, 277–292.
21. T. J. Horr, P. Arora, J. Ralston and R. St. C. Smart: 'XPS Film Thickness and Adsorption studies of CTAB and Silanes on Silica Plates', *Colloids and Surfaces*, submitted, 1994.
22. R. St. C. Smart: 'Surface Layers in Base Metal Sulphide Flotation', *Minerals Engineering*, 1991, **4**, 891–909.
23. K. Laajalehto, R. St. C. Smart, J. Ralston and E. Suoninen: 'STM and XPS Investigation of Reaction of Galena in Air', *Appl. Surf. Sci.*, 1992, **64**, 29–39.
24. R. St. C. Smart, P. Arora, R. Hayes, B.-S. Kin, C. Prestidge and J. Ralston: 'Chemical and Structural Alteration in the Surface Layers of Oxides and Sulphides', in *Science of Ceramic Interfaces II*, J. Nowotny, ed., *Materials Science Monographs 81*, Elsevier (Amsterdam), 1994, 311–340.
25. D. Fornasiero, F. Li, J. Ralston and R. St. C. Smart: 'Oxidation of Galena Surfaces: I. X-ray Photoelectron Spectroscopic and Dissolution Kinetic Studies', *J. Colloid Interface Sci.*, 1994, **164**, 333–344.
26. B.-S. Kim, R. A. Hayes, C. A. Prestidge, J. Ralston and R. St. C. Smart: 'Scanning Tunnelling Microscopy Studies of Galena. 1. The Mechanism of Oxidation in Air, *Applied Surface Science*, 1994, **78**, 385–397.

SURFACE FINISH IN MANUFACTURE: ITS ASSESSMENT AND SIGNIFICANCE

K. J. STOUT
School of Engineering, University of Huddersfield, Queens Gate, Huddersfield HD1 3DH
and
L. BLUNT
School of Manufacturing and Mechanical Engineering, University of Birmingham, UK

ABSTRACT

This paper demonstrates how 3D surface topography measurement is superior to conventional 2D characterisation in the assessment of surface texture of manufactured components. Examples are drawn primarily where engineering components are used in a tribological environment (friction and wear). The limitations of 2D assessment are highlighted in terms of sample significance and it is shown how 3D assessment provides a greater insight into the interactive properties of surfaces.

The paper is illustrated using visualisation techniques and numerical parameters (the 'S' parameters) as agreed at the 1992 and 1993 Brussels Workshops.

The scope and power of 3D analysis is discussed so that the delegate can gain an understanding of its potential for industrial application.

1. INTRODUCTION

The science of surface topography measurement began in the early 1930s when the first instruments were developed. The first measurement instruments provided a simple graphical output from which the peak to valley height information could be easily measured. This was rapidly followed by the first analogue instruments, developed in both the U.K. and in Germany, which enabled the average roughness parameter of surface profiles to be calculated. These instruments used simple 2CR filtering techniques to remove the longer wavelength (form and waviness) components. During the late 1960s the advent of digital computers gave the subject of surface analysis a significant boost, and for the first time digital techniques were applied to the science to enable a range of parameters describing surface roughess to be developed. These additional parameters were introduced in a random attempt to attach functional significance to the behaviour of surfaces in a range of operating environments. These uncoordinated activities led to a multiplication in the number of numerical parameters, many of which were found to have very little functional significance.

The first attempt at bringing together leading workers in surface analysis was the introduction in 1968 of the International Conference[1] held at the University of Oxford. The conference proceedings became the initial comprehensive text of surface finish. Much research and development work was undertaken in subsequent years and a new series of triennial international conferences was introduced in 1979. The fourth event in this series held in 1988 at the National Bureau of Standards in Washington DC, and more so at the most recent event held in Leicester in 1991,[2] made it quite clear that there had been a significant shift in the research and industrial interest in this field toward the analysis and experimental investigation of 3D surface measurement. It is therefore timely, due to the growing interest in 3D analysis, to formalise the procedures of 3D data capture, filtering and digital analysis in order to avoid the problems that occurred with 2D analysis, and the complications that have occurred in the attempt to specify leading international standards.

It has been found that the nature of the topography of machined and engineered surfaces has a significant effect on the behaviour of asperity contact. The extent of the interaction influences the life cycle of the functional activity of the components. This may in some cases be expressed in terms of wear and in other cases seen in terms of the effects of the inefficiency of the components on the environments. The contact behaviour of engineering surfaces has been the subject of much research. A critical factor in optimising the design of components is an understanding of the performance and wear characteristics of functional surfaces when in operation. The mechanism of wear does not always result in catastrophic failure; however, the overriding problems induced by wear reduce the efficiency of contacting surfaces, and can lead to increased power demands, increased lubricant consumption and vibration problems. To move towards a comprehensive understanding of surface wear it is necessary to study aspects of the material structure, the topographical shape, lubrication and the tribological properties of the surface.

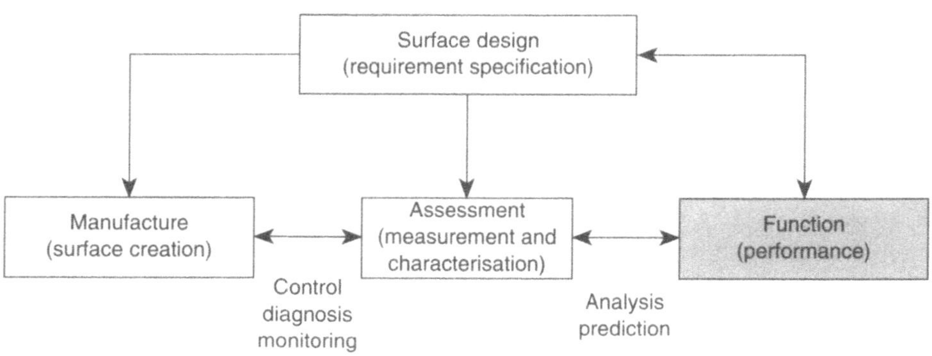

Fig. 1. The relationship between the surface design, function, manufacturing and assessment.

Engineering surfaces are created by a large variety of manufacturing processes. Sometimes they are left with the texture caused by the major operation which formed the bulk shape, e.g. turned, milled, rolled, lapped. Many surfaces, however, are designed to interact with other solids, and are often further treated, or surface engineered, to give a desired functional performance, a good example being the plateau honing of engine cylinder bores to provide a good lubricated bearing surface. The loop between surface design, manufacture assessment and functional performance is shown in Fig. 1.

Wear, friction and lubrication characteristics of a functional surface are intrinsically linked although they are rarely studied in tandem. There is no universal relationship between them; for example, small changes in friction may result in large changes in wear. Additionally changes in the wear mechanism may be indicated through sudden changes in friction. Wear and friction are not intrinsic properties of a material but rather the engineering system in which the materials operate.

1.1 Surface Creation

Engineering surfaces are created by a large variety of manufacturing processes. Sometimes they are left with the texture caused by the major operation which formed the bulk shape, e.g. turned, milled, rolled, lapped. Manufacturing problems can also play a significant role in the final surface shape. Peters and Vanherck have demonstrated that vibrations from electromotors can significantly influence the waviness from ground surfaces.[3] Gaudissart and Peters have also established a relationship between specimen roughness and the dynamic stiffness of the machine and Gaudissart and Vanherck have shown that form error also has a vibrational ancestry.[4,5] In another extensive study on the grinding process, a clear relationship has been established between the chip thickness (a function of the roughness) and cutting forces and wheel wear[6,7] and that it is possible to predict the most economical cutting conditions for the desired texture required.

Surfaces designed to interact with other solids are often further treated, or surface engineered, which affects only the surface layer, e.g. plating, painting. Thus, in new and developing technologies, the surface preparation is often aimed at producing a functional surface. The relationship between surface topography and its functional performance has been partially investigated using 2D surface analysis, but even this limited analysis is not yet fully understood and over the years the absence of clear direction has resulted in a proliferation of numerical parameters whose functional importance is questionable.[8] However, it is becoming increasingly obvious that a full understanding of the link between surface topography and functional performance can only be realised if a 3D approach to surface chartacterisation is used.

1.2 The 2D Approach

The most common technique for assessing surface characteristics is through the use of a stylus-based measurement system. The stylus is moved over the surface to be measured for a predetermined distance, or trace length. The vertical excursions of the stylus relative to a datum, when suitably magnified, is a measure of the deviations of the real surface from the geometrical surface within the calibrated tolerance range of the instrument. Early versions of these instruments used a valve amplifier and provided a single parameter reading from a meter. Later techniques used digital electronics to provide digitised trace data which could be analysed using a computer. This analysis provided several useful 2D characterisation parameters (average roughness R_a) RMS roughness (R_q), peak to valley roughness height (R_y), etc.) which have been incorporated into national and international standards.

The surface characteristics are quantified in terms of the ordinate height of surface asperities, i.e. by the amplitude parameters and their distribution, and the spacing of the asperities often in terms of their slope and correlation structure. The height ordinates are measured relative to a reference line which defines a datum. Ideally this line should follow the geometrical shape of the surface so that the topography can be measured as deviations from the line. It follows that, in 2D surface assessment, the technique used to define this mean line is critically important to the measuring system and, as such, suitable techniques are detailed in national and international standards. It is interesting to note that different standards define the datum line in different ways.[9,10]

In assessing different surfaces, the assessment length (cut-off value) and orientation of the trace is critically important. As a result of this, international standards specify cut-off lengths for a range of engineering surfaces. In digital surface characterisation the influence of sample spacing on subsequent parameters has yet to be fully investigated. The extent of variation in roughness parameters, obtained using 2D surface analysis procedures, may be accounted for by both.

(i) variations due to the practical limitations of the measuring system, and,
(ii) variations due to the surface structure.

Figure 2 clearly shows the limitation of 2D surface analysis. The two 2D traces are taken parallel and perpendicular to the lay of a bored surface and their height parameters are completely different and the structure of the surface is more difficult to envisage. 3D analysis on the other hand shows a true representation of the structure and functional inferences are more easily drawn from this true areal representation. Statistical analysis of 2D surface data and their distributions for surface profiles has produced extensive information relating to the character of manufactured surfaces. However, the inevitable conclusion from such analysis is that there is a fundamental limitation in using 2D analysis in an attempt to describe a 3D surface. Obviously, if the surface structure is anisotropic, then very significant variations in its height distribution will be obtained especially if profiles are traced in differing directions relative to the predominant lay.

1.3 3D Surface Analysis

Although a range of national and international standards exist for the assessment of surfaces in 2D,[11] the same is not true for 3D analysis. In order to establish a coherent approach to 3D surface topography analysis and communication it is necessary to have a recognised international standard available for use by industrialists and

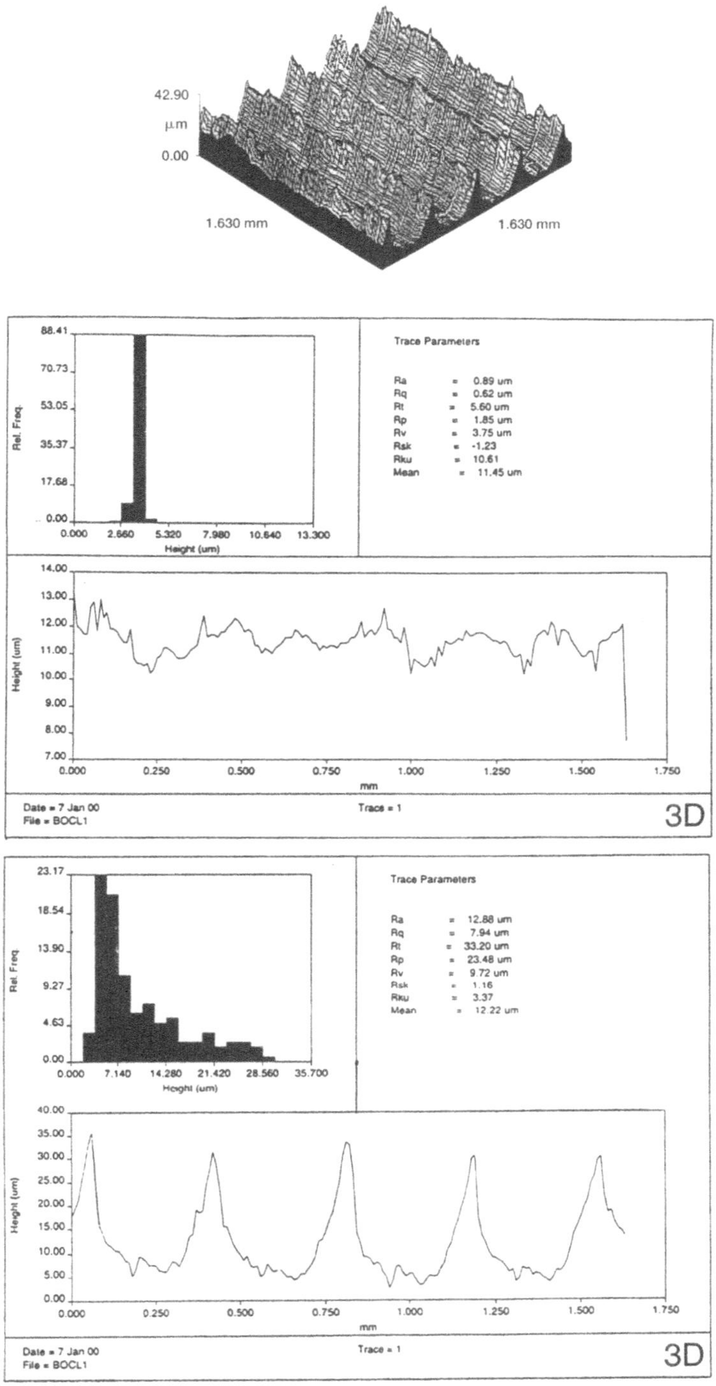

Fig. 2. 2D profiles and 3D map measured from a bored surface 2D and 3D.

researchers of the subject. To develop such a standard, it is necessary to investigate the variables which influence the measurement and analysis[12] (i.e. the fidelity). The most fundamental of these variables is described below.

In 2D analysis it was stressed that the determination of a suitable reference line is fundamental to the characterisation of the surface. In 3D analysis this requirement is extended to the establishment of a suitable reference, or datum, plane. Obviously, this plane should be sufficiently accurate to ensure that the height reference between successive profiles is recorded to an accuracy compatible with the smallest height measurement that can be made. Also, in recording multiple profile data, it is necessary to ensure a common origin at the start of each profile. This relocation must be made to an accuracy of at least the minimum distance between sample points.

Fundamental to determining the fidelity of a 3D surface analysis system is establishing the degree of precision to which the three ordinates can be measured. Digital techniques which are used to collect the data, such as the use of analogue to digital conversion, influence the accuracy of the acquired data. The minimum measurable distance that can be made using such digital conversion techniques is known as the quantisation level. The effect of the magnitude of the quantisation level on surface characterisation has yet to be fully understood in both 2D and 3D.

As a result of the largely uncoordinated development of 2D surface metrology well over 100 numerical roughness parameters were developed, with many of these having only limited significance.[8] To avoid the recurrence of this situation a structured European Community-based international programme has been initiated with the aim of the development of a new, fundamental, 3D surface measurement standard.[12] At the 1992 Brussels workshop organised by the E.C. and the University of Birmingham, Stout proposed that industrialists should agree upon a 'minimum set' of thirteen numerical surface characterisation parameters. He suggested that they be termed the 'S' parameters. After a period of discussion the fourteen parameters presented in Table 1 were tentatively approved. The document which followed[13] also proposed recommendations for measurement strategy, where the measurement cycle times are discussed and a maximum stylus speed of $0.5\ \mu ms^{-1}$ was recommended. Raster scanning with sampling areas of 128×128 or 256×256 orthogonally equally spaced data points was later accepted. A unified surface data file format was agreed. A least squares reference datum for 3D characterisation was agreed and has been defined as: 'a reference plane having the form of the geometrical surface and dividing the surface so that within the sampling area the sum of the squares of the surface departures from the plane is minimal'. It was suggested that electronic filters as recommended in 2D standards should not be used in a 3D measurement and any filtering, when it is necessary, should be conducted in the digital domain after the whole sampling area has been digitised. The arbitrary nature of the terms roughness waviness and form is such that they should not be used and instead the term surface data bandwidth should be applied when filtering is envisaged; Gaussian and zonal filters are recommended for specific applications. Autocorrelation, spectral moment analysis power spectral analysis and time series modelling are all discussed in terms of their application to surface texture analysis.

Table 1. 3-Dimensional roughness parameters

Amplitude parameters	
Sq	Root-mean square deviation of the surface from a plane
Sz	Ten point height of the surface
Ssk	Skewness of the surface
Sku	Kurtosis of the surface
Spatial parameters	
Sds	Density of summits of the surface
Str	Texture aspect ratio of the surface
Std	Texture direction of the surface
Sal	Fastest decay autocorrelation length
Hybrid parameters	
Sq	Root-mean slope of the surface
Ssc	Arithmetic mean summit curvature of the surface
Sdr	Developed surface area ratio
Functional parameters for characteristic bearing and oil retention properties	
Sbi	Surface bearing index
Sci	Core oil retention index
Svi	Valley oil retention index

2. 3D SURFACE ASSESSMENT

The following gives a number of examples of the use of the newly developed set of surface parameters.

Figure 3 shows the range and values of parameters obtained from a ground surface. The figure also shows the value of 3D representation of the surface measurement technique. Figure 4(a) (b) and (c) shows the development of the functional parameters across a number of surfaces. It is considered that the best way to characterise functional properties is to use indexes rather than actual physical quantities. For example, it is easier to envisage whether a functional property is favourable if an index is small or large rather than if it is given by an absolute value of, say, 1 or 5 μm.

These concepts are illustrated by considering the three surfaces shown in Fig. 4, where the isometric plots, height distributions and bearing area curves are shown. Figure 4(a) is a bored surface and clearly has poor bearing properties, while retaining a high oil retention volume. Figure 4(b) is an engine cylinder bore surface having much better bearing properties than the bored surface; however, the oil retention volume is somewhat smaller. Figure 4(c) is a plateau honed surface from an engine bore; this surface has good bearing properties with improved oil retention capability. The oil retention aspect of this surface is different to that of the bored surface in that the oil is retained in the valley zone as opposed to the core zone. These features are clearly illustrated from the bearing ratio curves.

If two horizontal lines are drawn on the bearing ratio curves as shown in Fig. 4,

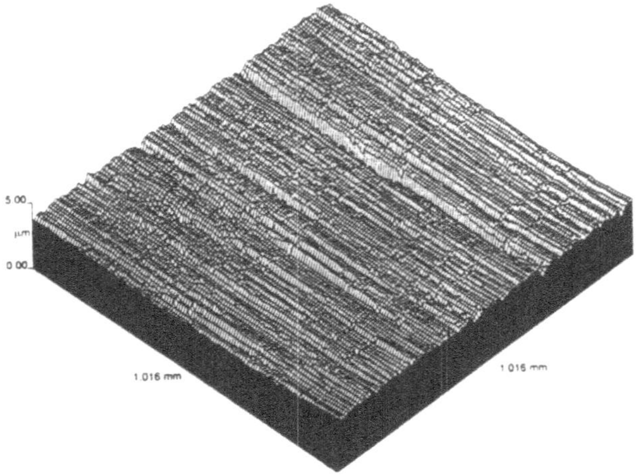

3D Surface Parameters

Sa = 0.12 mm, Sq = 0.15 mm,
Sz = 1.51 mm, Ssk = 0.70,
Sku = 5.88, Sds = 924.4 mm^{-2},
Sti = 0.053, Std = 0°,
SΔq = 0.025, Ssc = 0.0034,
Sdr = 0.030, Sbi = 0.74,
Sci = 1.24, Svi = 0.15.

Fig. 3. Ground surface and 3D parameters.

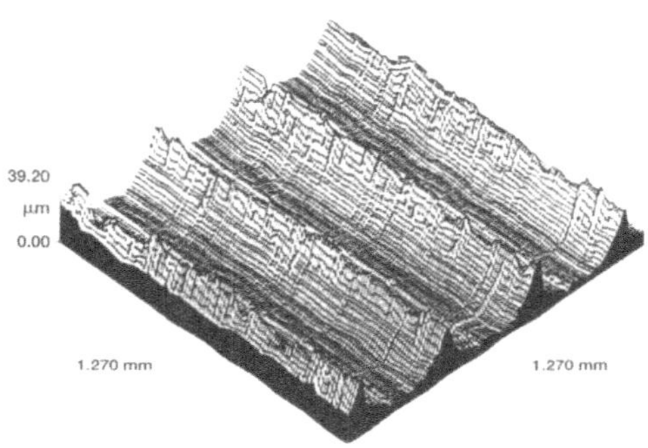

(a) Isometric plot of the surface

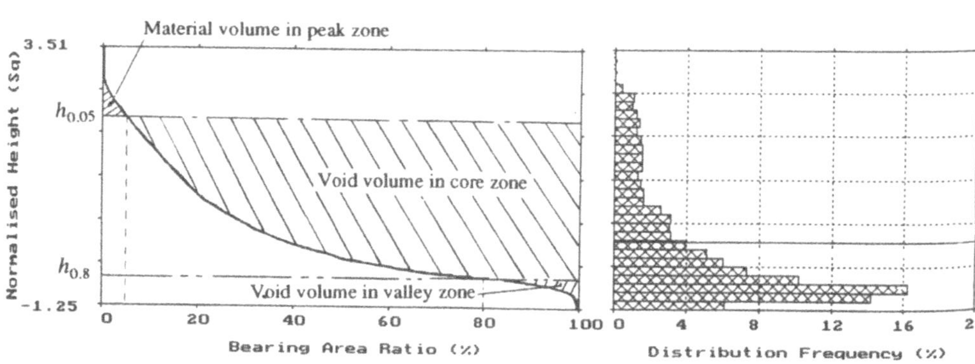

(b) Surface bearing area ratio

(c) Height distribution

Fig. 4a. Bearing ratio and height distribution of bored surface.

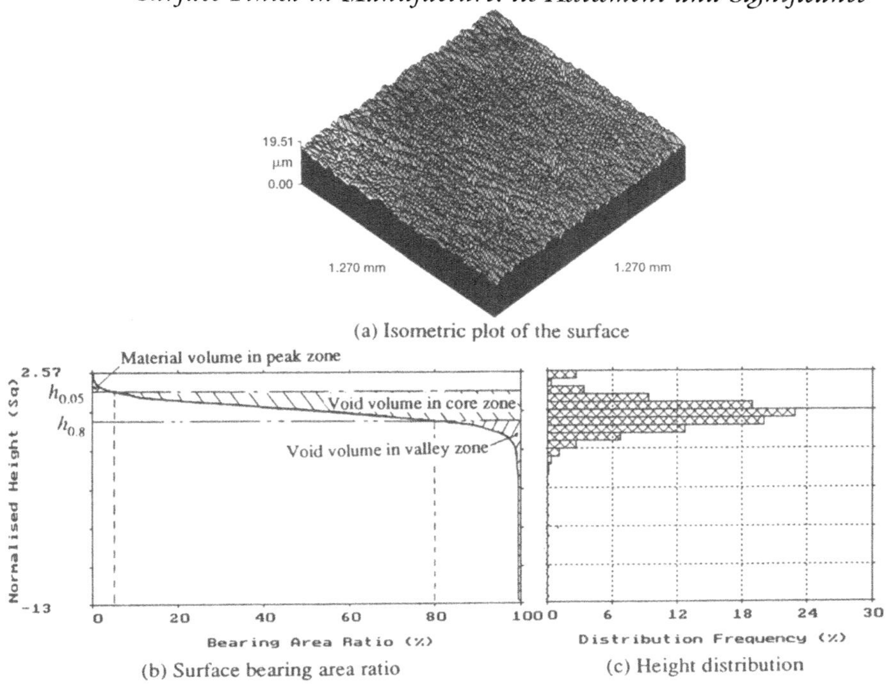

Fig 4b. Bearing ratio and height distribution of engine bore surface.

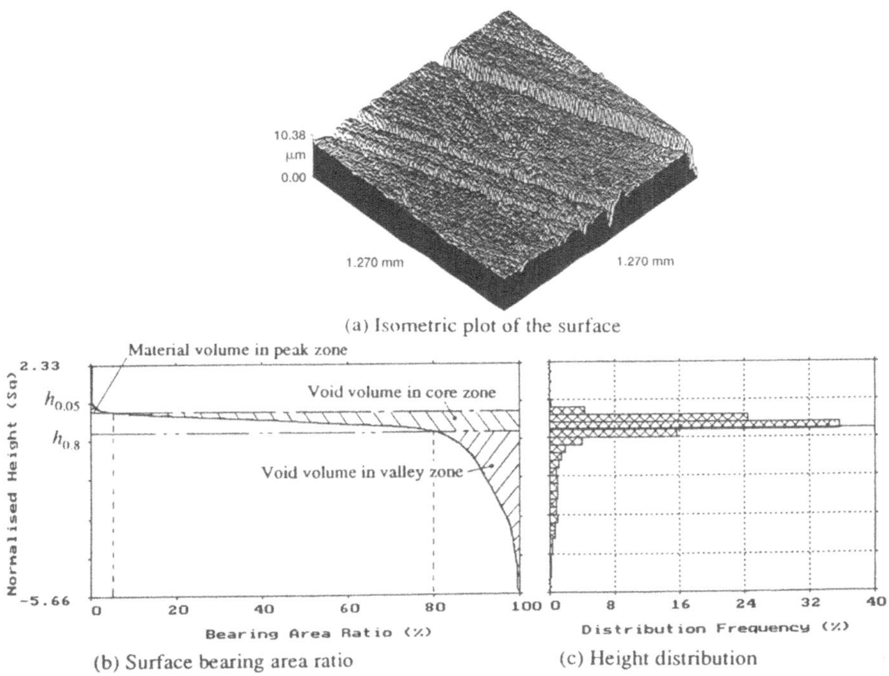

Fig 4c. Bearing ratio and height parameters of plateau honed surface.

one intersects the curve with 5% bearing area denoted, $h_{0.05}$, the other intersects with the 80% bearing area denoted, $h_{0.8}$. The curves are then classified into three zones similar to that denoted in DIN 4776, i.e. peak zone, core zone and valley zone. The material volume of the unit sampling area of each zone is actually the area enclosed beneath the surface bearing area ratio curve and above the bottom horizontal line of this zone. The void volume of the unit sampling area of each zone is the area enclosed beneath the top horizontal line of this zone and above the surface bearing area ratio curve. It can be seen clearly that bored surface has a larger void volume of the unit sampling area in the core zone but a smaller void volume of the unit sampling area in the valley zone. The plateau honed surface, however, has a comparatively smaller void volume of unit sampling area in the core zone and a comparatively larger unit void volume in the valley zone. For the engine bore surface the void volume of the unit sampling area in neither the peak or the valley zone is significant. These features illustrate that the oil retention volume property is most effective in different zones for the different surfaces. From Fig. 4 it can be seen that a good bearing surface has a fast increase of the bearing area ratio. If the volumes are normalised by division of the S_q this gives the function index parameters; for example, a high index indicates good bearing properties. The function index parameters are shown in Table 2.

Table 2. Functional indexes of surfaces in Fig. 4

	Sa/μm	S_{bi}	S_{ci}	S_{vi}
Bored	8.01	0.44	2.24	0.024
Engine bore	1.25	0.73	1.22	0.16
Plateau honed	1.32	1.38	0.42	0.30

- Surface roughness (3D equivalent of R_a) S_a/μm
- Surface bearing index S_{bi}
- Core oil retention index S_{ci}
- Valley oil retention index S_{vi}

The texture aspect ratio parameter S_{tr} has been developed in order to identify texture pattern, i.e. isotropy or anisotropy. The parameter is based upon the ratio of the slowest to the fastest decay of the autocorrelation function in all directions. A value of <0.3 indicates a strong anisotropic tendency associated with a strong texture, a value >0.5 indicates strong isotropy in all directions, as may be found from an E.D.M. surface (~ 0.8). The direction of surface texture lay can be identified using the S_{td} parameter. This parameter is based upon the angular spectra of an auto power spectral density, APSD, analysis of surface. The direction of the texture is taken as the position where the maximum angular spectrum appears (when the measurement trace direction is perpendicular to the lay the lay direction angle is 0°). Clearly the S_{td} parameter is meaningless if the texture aspect ratio, S_{tr} is >0.5. An example of the two above parameters is given in Fig. 5 and Table 3.

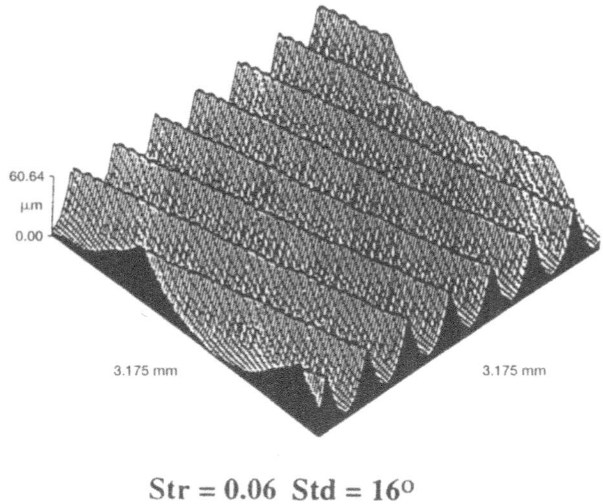

Fig. 5. Shaped surface used for showing texture index and texture direction.

Table 3. Texture parameters for a shaped surface

Surface	S_{tr}	S_{tdo}
Shaped	0.06	−16

The texture aspect ratio for a number of differently manufactured surfaces is shown in Fig. 6.

Of the new hybrid parameters the S_{dr} parameter developed surface area ratio is particularly useful for analysing the interfacial areas changes brought about by a

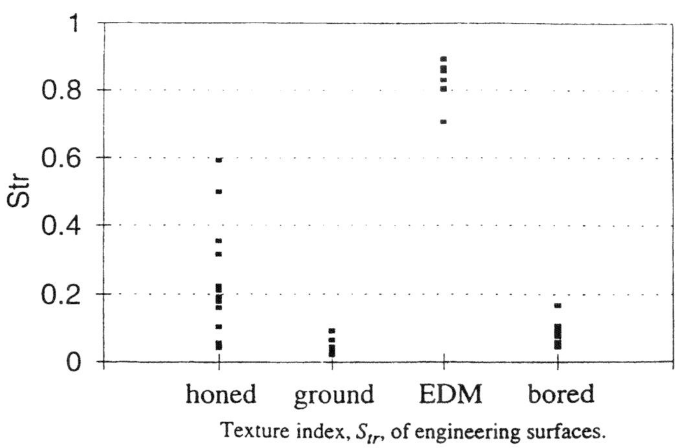

Fig. 6. Texture index of a number of manufactured surfaces.

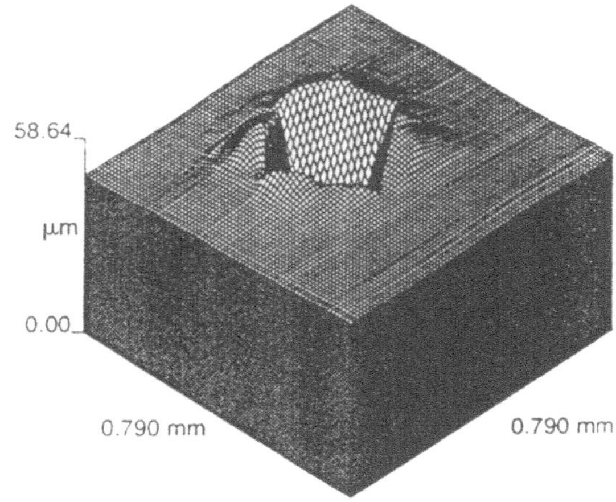

Fig. 7. Figure showing a 3D representation of a microhardness indentation.

particular surface modification. The parameter is based upon the ratio of the sampled area to the real surface area. Figure 7 shows the increase in the surface area brought about by a hardness indentation, in this case the surface area increase/ratio has the value of 0.947. A low value of this parameter indicates a smooth surface, e.g. polished has a ratio of 0.0086 compared to a ground surface with a ratio of 0.0314. The parameter allows the analysis of the effects of surfaces modification operations to be carried out.

3. DISCUSSION AND CONCLUSIONS

There is now a well defined and accepted means of characterising the topography of 3D surfaces. This integrated approach to characterisation has been accepted in a second workshop held in Brussels in June 1993 under the auspices of the E.C. The fact the workshop, at which the attendees were mostly industrialists, led to the acceptance of the parameter set described and defined here, presented the first steps to an acceptable method of international communication of 3D topography. The authors do not suggest that the parameters recommended and accepted at that meeting are sufficient to fully describe all forms of surface interaction but they provide a useful starting point for international communication and decision making.

As the subject of 3D characterisation and assessment gains increased usage by the industrial community it is anticipated that selected additional parameters will be added where appropriate, but it is believed that the current limited set of thirteen parameters provides the basis of any future extended set. Some of the extended parameters which may be introduced should be introduced as functional additions and should not cloud the issue of the basic descriptors described here.

REFERENCES

1. Proc. Int Conf., 'The Properties and Metrology of Engineering Surface', Oxford, published by Institute of Mechanical Engineering, Vol. 182, part 3K, 1967–8.
2. Proc. Int Conf., 'The Properties and Metrology of Engineering Surfaces', Leicester 1991, published by *Int. J. Machine Tool Design and Manufacture*, 1992.
3. J. PETERS and P. VANHERCK: 'The Unbalance of Electromotors and Their Influence on Surface Geometry in Grinding', *Annals of CIRP*, 1971, **xviv**, 585–589.
4. P. GAUDISSART and J. PETERS: 'Influence des vibrations sur le fini de surface et l'erreur de forme des passe de finition', *CRIF*, Institut de Mechanique, Heverlee, 1964.
5. P. GAUDISSART and J. PETERS: 'Relation entre la rugosite d'une piece tournee et la rigidite dynamique de la machine', *CIRP Annalen*, 1963, **xii**.
6. R. SNOEYS: 'The Mean Underformed Chip Thickness as Abasic Parameter in Grinding', *Annals of CIRP*, 1971, **20/2**.
7. R. SNOEYS, J. PETERS and A. DECNEUT: 'The Significance of Chip Thickness in Grinding', *Annals of CIRP*, 1974, **23/2**.
8. D. J. WHITEHOUSE: 'Parameter Rash – is there a cure?', *Wear*, 1982, **83**.
9. *BSI Assessment of Surface Texture, Part 1. Methods and Instrumentation*, British Standards BS 1134, 1988.
10. *ISO, Surface Roughness – Terminology – Part 1: Surface and its parameters*, International Standard ISO 4287/1, 1984.
11. *ANSI Surface Texture: Surface Roughness, Waviness and Lay*, American Standard ANSI B.46.1, 1985.
12. Commission of the European Communities, 'Development of a Method for the Characterisation of Surfaces in 3 Dimensions', Contract No. 3374/1/0/170/90/2/-BCR-UK 30.
13. K. J. STOUT, P. J. SULLIVAN, W. P. WONG, E. MAINSAH, N. LUO, T. MATHIA and ZAHOUANI: *The Development of Methods for the Characterisation of Roughness in Three Dimensions*, Published by the Commission of the European Communities.

RESIDUAL STRESS MEASUREMENT – METHODS, LIMITATIONS AND SIGNIFICANCE

D. W. BORLAND

Department of Materials Engineering, Monash University, Clayton, Victoria 3168, Australia

ABSTRACT

Methods for the measurement of residual stress are reviewed, especially in terms of their applicability to surface engineering processes. Hole drilling and X-ray diffraction techniques provide attractive combinations of accuracy and spatial resolution, while ultrasonic and magnetic methods are convenient in use despite difficulties in interpretation of their results.

1. INTRODUCTION

Residual stress is a phenomenon that is obviously important and yet very rarely allowed for in a quantitative way in the design of components or processes. We know that local residual stress will add to the stress imposed in service and that the combination may well lead to unsatisfactory performance, either during manufacture, e.g. distortion, cracking, or in service, e.g. stress-corrosion, fatigue. So there is a need to understand how residual stresses arise and how they may be measured and controlled.

In surface engineering there is, almost by definition, a strong probability that residual stress may exist, both in the surface layer and also at depth, the depth of particular importance being that where separation of the surface-treated layer from the bulk may occur, i.e. spalling. The origin of this stress may be a thermal or thermochemical treatment, as in induction hardening or carburising, or a result of the epitaxial growth of a surface layer, as in electroplating and some vapour deposition techniques. Many other processes in surface engineering may give rise to residual stress by combinations of thermal and compositional effects. These processes will be discussed in other sessions of this conference. In this paper we will not consider the origin of residual stress but concentrate on the means available for its measurement and the significance of such measurements for the practical applications of surface engineering technology.

As in many other areas of materials science, indeed of science in general, there have been great changes in procedures in recent years as a result of modern microchip technology, which allows the acquisition of data and its processing into user-friendly form with far less expenditure of time and effort than formerly. The principles involved in most, if not all, of the techniques for the measurement of residual stress have been well established for many years, but it is only recently that some of these techniques have become possible on a commercial scale.

2. TYPES OF RESIDUAL STRESS

It is now generally agreed that three types of residual stress should be distinguished.

- Type I stresses, otherwise referred to as macrostresses, are those that are developed on a scale larger than the grain size of the material.
- Type II stresses are those that vary on the scale of an individual grain. Such stresses may be expected to exist in multi-phase materials because of the different properties of the phases and in single-phase materials because of anisotropy in the behaviour of each grain.
- Type III stresses are those that exist within a grain, essentially as a result of the presence of dislocations and other crystalline defects. Types II and III are often grouped together as micro-stresses.

Most techniques for residual stress measurement yield results for the macrostress, i.e. Type I, because they average over many grains. However, Type II stresses may be important in service because, depending on the grain size, such stresses may contribute to failure, e.g. by delamination or spalling. Type III stresses certainly contribute to the finer details of mechanical behaviour and even to such large-scale phenomena as stress-corrosion but are generally ignored in discussions of residual stress because there is no useful way of allowing for their effects on the grosser features of mechanical behaviour that are usually of interest. In the following discussion we will distinguish the capability of various techniques for the detection and measurement of stresses of Types I and II, and follow traditional practice in ignoring the stresses of Type III.

3. MEASUREMENT OF RESIDUAL STRESS

Of the many techniques for the measurement of residual stress we may distinguish firstly those that require destruction of the component under investigation from those that do not. Obviously, destructive methods are only appropriate during the development of a process or for quality control by the inclusion of test specimens among the components being treated. On the other hand, non-destructive tests allow a program for quality control to be established on the basis of testing a sample from a batch, or even 100% inspection.

3.1 Destructive Methods

Because they make fewer demands on theory and require no significant approximations apart from the standard assumptions of classical elasticity, methods based on the removal of material from the body under investigation have great appeal. Relaxation as a result of this removal causes elastic strain, which can be measured by a variety of techniques.

There have been many methods of this kind used, the original one being the Sachs boring technique,[1] which is applicable to cylindrical objects and relies on the radial symmetry to be expected in the stress field remaining in a cylinder after boring out a layer from the inner diameter. Nowadays the most widely used of these techniques are variants on the theme of drilling a hole in the component.[2,3] The hole is usually quite small, often less than 2 mm in diameter, so the method might be classified as 'semi-destructive'. A strain-gauge rosette is attached to the surface, from which material is then removed – either a cylindrical annulus surrounding the gauges (the trepan or ring-core method) or a cylindrical hole at the centre of the rosette (the centre-hole method). The measured strains can be used to calculate the components of the original biaxial stress field at the surface, provided that no plastic flow occurs during the relaxation.

Unless the hole passes completely through a thin plate, it is not possible to relate strain measurements to stress solely by using the theory of elasticity. For the more interesting cases of blind holes in components of arbitrary shape, experimental calibration is highly desirable though calibration curves for particular rosettes and hole diameters are available.[2] The results obtained are not only the surface stresses but also stresses at a depth up to half the diameter of the hole, provided that strain measurements are made periodically as the hole is deepened.[3] The extensive literature describing these techniques, dealing with validation, calibration and methods of calculation has recently been reviewed by Procter and Beaney.[2]

While these techniques are soundly based in principle, there are experimental difficulties. The most important is the requirement that the drilling of the hole should not introduce any extra stress. There is plenty of evidence that conventional mechanical methods such as drilling or trepanning can introduce extra stresses, while non-mechanical methods such as electropolishing or etching do not produce holes of the required shape (i.e. of uniform diameter and flat bottomed). The most promising procedure is ultra-high-speed drilling using the type of drill that has been developed for dental applications; a conventional cutter is used with low rates of feed.[4] Air-abrasive jets have also been used with considerable success.[2]

The precision of the best hole-drilling techniques, under favourable conditions, is of the order of 5%.[2] The procedure is applicable to any accessible surface but requires considerable time, an hour or more, since strain gauges must be attached to the surface before cutting is begun. The stresses measured are usually those of Type I, although in extremely coarse-grained materials a small (say 1 mm diameter) hole might sample the stress in an individual grain.

3.2. Non-destructive Methods

3.2.1 X-ray Diffraction

Overwhelmingly the most highly developed and the most written-about non-destructive method is the use of X-ray diffraction. The basic principle is illustrated in Fig. 1. An X-ray beam normal to the surface is used to measure the interplanar

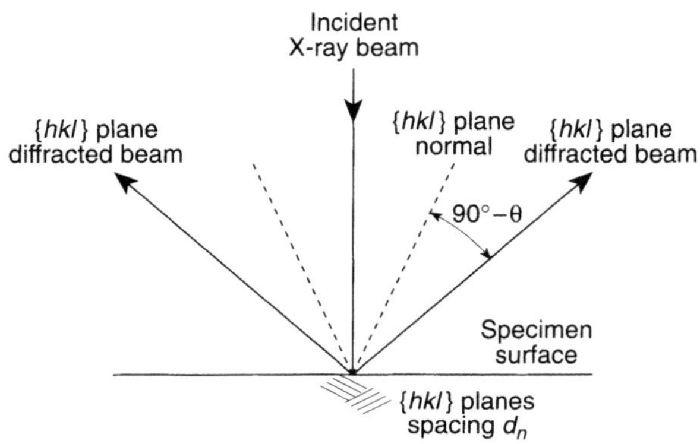

Fig. 1. Schematic geometry for the normal-incidence exposure in the X-ray diffraction method. Diffracted beams emerge from the surface along the generators of a cone (shown in section here). θ is the Bragg angle for diffraction from {hkl} planes, spacing d_n.

spacing of a suitable family of planes, {*hkl*}. In the presence of stress this spacing, d_n, differs from that in the unstressed state, d_o, so these planes can be used as a strain gauge if the unstressed spacing is known. This simple technique yields the sum of the two principal stresses in the surface, it being assumed that the third principal stress, normal to this free surface, is zero.

The technique is extended by taking another measurement of the {*hkl*} spacing, d_ψ, with the X-ray beam inclined at an angle ψ from the surface normal. When combined with the measurement at normal incidence the following relationship can be derived[5] from standard elasticity theory:

$$\sigma_\phi = [E/(1 + \upsilon)]\,[d_\psi - d_n)/d_o]/\sin^2\psi \quad (1)$$

where σ_ϕ is the normal stress in the direction defined by the angle ϕ (Fig. 2), E and υ are Young's modulus and Poisson's ratio. It should be noted that the members of the {*hkl*} family of planes that contribute to the diffracted beam at normal incidence are not the same as those that contribute when the incident beam is at an angle ψ.

There is an acceptably small change in accuracy if d_o in the above expression is replaced by d_n:

$$\sigma_\phi = [E/(1 + \upsilon)]\,[d_\psi - d_n)/d_n]/\sin^2\psi \quad (2)$$

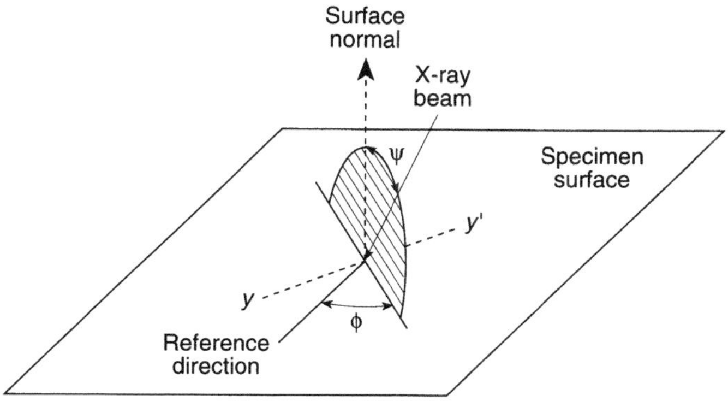

Fig. 2. The angles ϕ, ψ as defined for the two-tilt X-ray method. Note that the oblique angle of incidence, ψ, may easily be achieved by maintaining the direction of the X-ray beam constant and rotating the specimen about the direction *y–y'* in its surface.

and thus the normal stress in a direction at an angle ϕ, measured with respect to some arbitrary direction in the surface of the specimen, may be determined from two exposures. This is known as the 'two-tilt' method.[6] If the measurement is repeated for two further values of ϕ the complete biaxial stress state in the surface may be deduced.

A development from the two-tilt technique that has proved to be especially powerful is the '$\sin^2\psi$' method. According to equation (2) the stress σ_ϕ may be obtained as the slope of a plot of $(d_\psi - d_n)/d_n$ against $\sin^2\psi$ if a range of values of ψ is used. Linear regression improves the statistical accuracy of the determination but, more importantly, significant additional information may be obtained from the nature of the plot. It has been shown that characteristic departures from linearity are related to details of the stress field, the most important being due to steep stress gradients normal to the surface (ψ-splitting, Fig. 3(b)) and the existence of texture

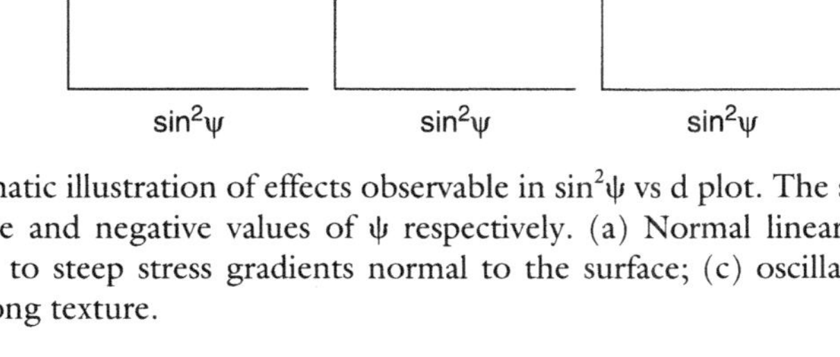

Fig. 3. Schematic illustration of effects observable in $\sin^2\psi$ vs d plot. The symbols **o** and **x** refer to positive and negative values of ψ respectively. (a) Normal linear behaviour; (b) ψ-splitting due to steep stress gradients normal to the surface; (c) oscillation due to the presence of strong texture.

(i.e. preferred orientation) in the crystalline structure (oscillations, Fig. 3(c)). There has been extensive discussion in the literature of these effects; reviews by James and Cohen[6] and more recently by Hauk and Niku-Lari[7] explore the issues in detail.

Precision in the measurement of d-spacings is obviously necessary. While the required precision can easily be achieved in the laboratory it is not such a simple matter when attempts are made to minimise the time required for a measurement and to make the apparatus readily portable. Sophisticated procedures for the analysis of data are now readily available to deal with the first of these difficulties, while the development of position-sensitive proportional detectors and miniature X-ray tubes has contributed substantially, at some cost, to the improvement of portability.[6,8] A discussion of these developments is given by James and Cohen[6] who have shown that stress can be measured to an accuracy of ±35 MPa in no more than 20 s, using an apparatus that weighs 7–11 kg and is readily portable.

There is another significant problem in the determination of stress by X-ray diffraction that needs to be borne in mind. It is worth emphasising because the point is potentially applicable to all other stress measurement techniques: what we measure is *strain*. The stress that causes this strain is inferred by the use of elasticity theory. In the X-ray diffraction case, equations (1) and (2) are based on the assumption of isotropic elasticity, and many measurements of residual stress have been carried out using this assumption. But individual crystals are anisotropic; the assumption of isotropy is based on measurements involving many crystals, assumed to have random orientations. While X-ray measurements usually extend over many grains, the planes that contribute to the measured part of the diffraction pattern are highly selected, i.e. only those that are suitably oriented, given the chosen direction of the incident beam. There has been debate about the use of more suitable elastic constants to achieve more accurate values of stress; theoretical and experimental techniques for obtaining more appropriate values of the elastic constants have been proposed but there is no widely accepted single procedure for dealing with this problem.[7]

Finally, it must be remembered that X-ray procedures measure the stress in the surface. Stress gradients perpendicular to the surface may be detected if the $\sin^2\psi$ technique is used, but there is no readily available way to measure this gradient from measurements at the surface and, of course, no information about stresses, which are almost certainly triaxial, at depths greater than the penetration distance, typically a few μm. Thus depth profiling of the residual stress pattern must be undertaken by removal of the surface layer followed by measurement on the new free surface, with adequate allowance for the relaxation of stress resulting from removal of a layer. Removal is usually done by electropolishing, but this technique does not produce an adequately flat surface if more than a few μm are removed.[9]

3.2.2 Ultrasonic Methods

The propagation of sound waves through a material depends on its elastic behaviour. In principle, therefore, since residual stress changes the distance between atoms the elastic behaviour is modified and therefore the presence of residual stress is detectable by ultrasonic techniques. In practice there are substantial difficulties, the most important being that the effect of stress on the propagation velocity is generally smaller than the effects of texture and microstructure.[10] The spatial resolution of all but longitudinal focused probes is also inadequate for many applications.[10,11]

It has been shown[12] that the textural problem can, in theory, be overcome if surface-skimming shear horizontal (SH) waves are used, propagating in one of the directions of principal stress in the surface and polarised parallel to the other principal stress direction. By rotation of the probe through 90° the directions of propagation and polarisation can be interchanged; then the difference between the two velocities of propagation is directly related to stress by the equation

$$\sigma_i = 2G[(V_{ik} - V_{ki})/V_{ik}] \tag{3}$$

where σ_i is the stress in the direction i, G is the shear modulus and V_{ik} the velocity of a SH wave propagating in the i direction and polarised in the k direction.

Measurements of residual stress have also been made[7] by comparing the velocities of propagation of an ordinary transverse wave pulse when it is polarized in two mutually perpendicular directions. The times of flight, t, from the injection of the pulse to its reception after reflection from another surface, are given by the formula

$$(t_{ij} - t_{ik})/t_{ik} = (\sigma_k - \sigma_j)(4G + n)/8G^2 \tag{4}$$

where i, j, k are the principal stress axes and n is one of the third-order elastic constants.[13] Note that this formula does not contain the propagation distance; thus the stress measurement does not depend on accurate knowledge of the distance travelled by the waves. There are, however, questions of whether the two waves travel exactly the same distance.[14] This procedure does not eliminate the problem of texture, and gives an average stress over the distance of propagation provided that the stress does not vary greatly over a distance of one wavelength (of the order of 1 mm for waves of frequency 1 MHz).

These ultrasonic methods have been made possible by the very sophisticated electronic apparatus that is available today and by the development of sophisticated transducers. Their attraction lies in the portability of the apparatus and the speed with which measurements can be made. While successful applications have been reported, it is difficult to evaluate the applicability of ultrasonic methods to the routine determination of residual stress, especially in applications to surface engineering where stresses in thin surface layers are usually of considerable interest. Rayleigh waves should be particularly useful, since they propagate in the surface layer.

An interesting study has been reported by Todaro *et al.*,[15] who investigated residual stresses in a long, thick-walled cylinder after quenching. This cylinder, presumably a gun-barrel forging, was 239 mm in outside diameter and 94 mm inside diameter. After quenching to 93°C in 12 minutes (and, presumably, air-cooling from this temperature) a 75 mm thick slice was cut, 300 mm from one end. The slice was then ground on each end to give a specimen 50 mm in length. Ultrasonic measurements of stress were made from one end along two diameters as shown in Fig. 4, then X-ray measurements were made on this face after removal of 0.1 mm by

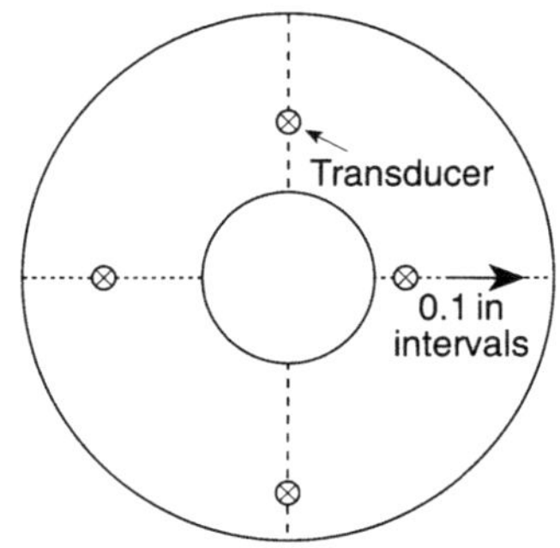

Fig. 4. Scheme of transducer locations for the ultrasonic measurements of residual stress reported by Todaro *et al.*[15]

electropolishing. The ultrasonic method involved transverse waves polarised in two directions at right angles, as described above, but in this case the velocities of the waves were measured directly rather than the difference in velocity between waves of different polarisation. Some results from this study are shown in Fig. 5, in which the tangential stress is plotted against radial position for three determinations – ultrasonic, X-ray using a single-exposure oblique-incidence technique and portable equipment and X-ray using the $\sin^2\psi$ technique with a conventional goniometer. It should be noted that the measured stress for the ultrasonic technique is an average over the 50 mm thickness of the specimen, while the X-ray measurements refer only to the surface layer.

The results shown in Fig. 5 seem to illustrate two important points. In all three sets of measurements the qualitative stress distribution is the same, i.e. compressive

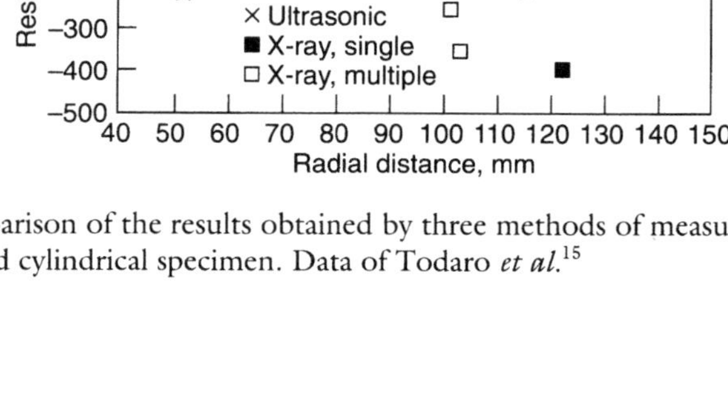

Fig. 5. Comparison of the results obtained by three methods of measuring residual stress in a thick-walled cylindrical specimen. Data of Todaro *et al.*[15]

stress at the inner and outer surfaces, with a peak of tensile stress at an intermediate diameter. Comparison of the results from the ultrasonic technique with those from X-ray diffraction shows that the radial position of the peak tensile stress is markedly different; this is probably due to unreliability in the spatial resolution of the ultrasonic measurement, though this point is not commented on by the authors. The other point of interest refers to the X-ray methods: the difference in the scatter for the two sets of measurements is some indication of the relative accuracy of the two X-ray techniques.

It is evident that ultrasonic methods of residual stress determination still require development before they can provide quantitative determinations of residual stress of good spatial resolution and adequate accuracy. On the other hand, the potential ease of use and portability of ultrasonic systems suggests that further development is likely and that improvements, particularly in transducers for the generation and detection of suitable ultrasonic impulses, will before long result in readily available instruments for general application. It is likely, however, that expert operators will still be required because it has never been an easy matter to ensure the reliable generation, reception and interpretation of ultrasonic signals.

3.2.3 Magnetic Methods

In ferromagnetic materials changes in the magnetic field strength produce changes in magnetic induction through the movement of domain walls between regions whose magnetisation is in different directions. These domain walls (Bloch walls) do not move smoothly and continuously as the magnetic field strength changes, but in irregular jumps, each of which produces an electromagnetic signal that can be detected by a search coil placed close to the material (Fig. 6). This is the Barkhausen effect, first described in 1919.[16] In magnetostrictive materials an acoustic effect may also be observed – magneto-acoustic emission.[17]

The strength of the Barkhausen noise signal is affected by the microstructure and also by the state of stress of the material under investigation and thus, in principle, it is possible to detect and even to measure the level of residual stress by detection of signals generated by the movement of these Bloch walls.[17,18] Calibration is required to relate the amount of Barkhausen noise to the level of stress, and this must be done using material with a similar, preferably identical, microstructure.

The spatial resolution attainable depends on the volume of material that is magnetised. Since the depth of penetration of the magnetic field depends on its frequency, the technique is potentially very useful for the study of near-surface regions such as those generated by many surface engineering treatments. On the other hand it is difficult to separate microstructural and textural effects from those due to residual stress.

The technique still requires much development if it is to be used for quantitative stress measurements, but the fact that it is potentially usable with readily portable equipment that generates no harmful radiation and requires only a few seconds for each measurement suggests that further development to overcome the problems is likely.

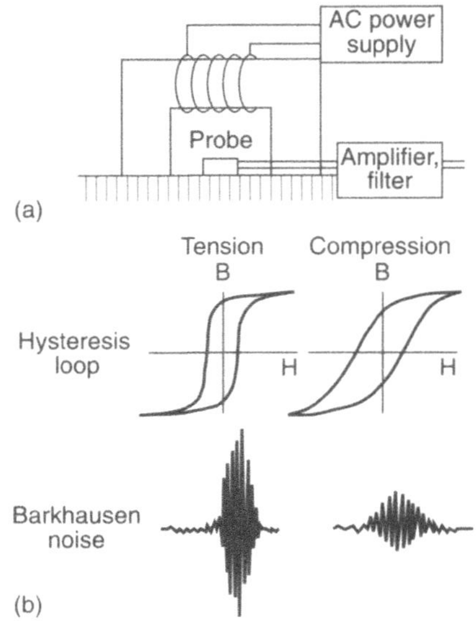

Fig. 6. Barkhausen noise analysis. (a) Schematic arrangement of magnetising and detection systems; (b) relationship between the hysteresis loop and Barkhausen noise for samples in tension and compression (from Tiitto[18]).

4. SIGNIFICANCE

It is evident that several methods are available for the measurement of residual stress. They have varying degrees of accuracy and convenience, but the best-developed of them can measure stress with impressive precision though sometimes they may involve considerable time and effort. It will be appropriate to conclude with a comparative assessment of the applicability of each technique, but before doing so it will be useful to consider the purpose to be served by the measurement.

There are three broad purposes for the measurement of residual stress.

- For the comparison of model calculations with stresses that are generated in practice. The development of models, and calculations based on these models, is an extremely popular activity these days, especially now that finite-element and other methods are available and nearly every desk carries a computer that is capable of carrying out the calculations. But our own knowledge of the real world and how it works is still not detailed enough for these models to be used with confidence unless they have been adequately tested. While, as discussed below, there may not be any pressing need in industrial applications to know precise values of residual stresses there is a much more rigorous requirement for models to be tested by the generation of accurate experimental data.
- For the design of components. In the best of all possible worlds, the designer would include in a quantitative way the residual stress generated during manu-

facturing in an estimate of the stress that a component will experience in service. In our real world, while the designer knows, for instance, that residual compressive stress will be generated in a component because of a specified surface treatment such as carburising or induction hardening, this is not included specifically in design calculations. The presence of such stress is assumed, even allowed for, in design codes, but its magnitude is not considered. Thus, for design purposes the interest is in whether stresses at a particular point, usually on the surface, are compressive or tensile, not on their absolute values or even, in many cases, whether they are large or small.

- For quality assurance in manufacture. Once again, absolute values of residual stress are not of interest, after the manufacturing procedure has been validated and permissible variations have been decided. It is not even necessary to measure residual stress itself, provided that some quantity that is known to be related to residual stress can be measured with sufficient precision for the purposes of quality control.

From the above we can see that, for most applications, precision is not necessarily the most important factor to consider in choosing a method for measurement of residual stress. Speed, ease of use and, above all, cost, are the important attributes for a production system. Designers, likewise, do not wish to know what the residual stress is likely to be to the last MPa, but they are interested in the avoidance of dangerous tensile stresses at surfaces where fatigue cracks might initiate. It is really only in the modelling and development of manufacturing processes that more accurate values are required and in most of these cases speed, convenience and portability are of secondary importance.

With these considerations in mind, the methods described above may be characterised as follows.

Destructive methods are slow, but potentially very accurate in the measurement of Type I stresses. With careful attention to the problem of introducing new stresses during the removal of material, accuracy of better than 10% should normally be achieved. The equipment is portable and suitable for use in the field. Modern hole-drilling techniques may often remove so little material that the technique is effectively non-destructive when applied to the measurement of stresses close to or at the surface.

X-ray methods are very accurate in the laboratory, less so if portable equipment is used in the field. If measurements below the surface are required the method is tedious and needs considerable care to achieve reliable results. Modern miniaturised equipment is suitable for use in the field, though this equipment is relatively costly and there are reservations these days about the use of ionising radiations in locations that are not strictly controlled. The stresses measured by X-ray diffraction include Type II as well as Type I. Problems may be encountered if strong texture is present in the material.

Ultrasonic methods are not capable of the same precision or spatial resolution as X-ray methods, but are more easily and quickly made and, for measurement in the field, it is an advantage that they do not require possibly dangerous radiations. Microstructural and textural effects on the propagation of ultrasonic waves may mask the effects of residual stress and there are difficulties in separating meaningful signals from the inevitable noise that accompanies measurements, especially in the field. These techniques are, perhaps, best suited to qualitative measurements but their potential is nevertheless great because of their convenience and portability. Improvements in the reliability, accuracy and spatial resolution may be expected.

Magnetic methods, in their application to ferromagnetic materials, have similar operational characteristics to those of ultrasonic methods, in that measurements are easily and quickly made in the field. Although the physical processes underlying the technique are different from those associated with ultrasonic methods, there are similar problems in separating the effects of stress from those of microstructure and texture. Despite these problems it is likely that the techniques will be developed further because of their operational advantages.

5. CONCLUSION

Of the many potential methods for the measurement of residual stress, the traditional techniques of material removal and X-ray diffraction analysis are those most capable of accuracy and good spatial resolution, especially when applied to the measurement of surface and near-surface stresses. Developments in hole-drilling techniques have now established them as pre-eminent for routine purposes, both in the laboratory and in the field, though the time required for a measurement is still, inevitably, much longer than that for other techniques. Modern X-ray equipment is much more portable and convenient than formerly, and residual stress measurements of adequate accuracy may now be made quickly in the field as well as in the laboratory. Ultrasonic and magnetic methods show great promise in their convenience of use, but there is room for their further development in terms of spatial resolution and accuracy.

ACKNOWLEDGEMENTS

I am grateful to Mr R. A. Coyle for helpful discussions, and to Professor P. L. Rossiter for the provision of facilities at Monash University.

REFERENCES

1. G. Sachs: *Z. Metalk.*, 1927, **19**, 352.
2. E. Procter and E. M. Beaney: *Advances in Surface Treatments, Vol. 4 Residual Stresses*, A. Niku-Lari ed., Pergamon, Oxford, 1987, 165.

3. A. Niku-Lari, J. Lu and J. F. Flavenot: *Advances in Surface Treatments, Vol. 4 Residual Stresses*, A. Niku-Lari ed., Pergamon, Oxford, 1987, 199.
4. J. M. Boag, M. T. Flaman and J. A. Herring: *Residual Stress in Design, Process and Materials Selection*, W. B. Young, ed., ASM Int., 1987, 1.
5. H. P. Klug and L. E. Alexander: *X-ray Diffraction Procedures*, Wiley, New York, 1974, 347.
6. M. R. James and J. B. Cohen: *Treatise on Materials Science and Technology, Vol 19 Experimental Methods Part A*, H. Herman, ed., Academic Press, New York, 1980, 1.
7. V. Hauk: *Advances in Surface Treatments, Vol. 4 Residual Stresses*, A. Niku-Lari, ed., Pergamon, Oxford, 1987, 251.
8. C. O. Ruud and D. J. Snoha: *Residual Stress in Design, Process and Materials Selection*, W. B. Young ed., ASM Int., 1987, 151.
9. D. L. Milam: *Residual Stress in Design, Process and Materials Selection*, W. B. Young, ed., ASM Int., 1987, 77.
10. G. C. Johnson and J. J. Dike: *Review of Progress in Quantitative Nondestructive Evaluation, Volume 7B*, D. O. Thompson and D. E. Chimenti, eds, Plenum Press, New York, 1988, 1391.
11. D. T. MacLauchlan, L. R. Burns and G. A. Alers: *Review of Progress in Quantitative Nondestructive Evaluation, Volume 7B*, D. O. Thompson and D. E. Chimenti, eds, Plenum Press, New York, 1988, 1399.
12. R. B. Thompson, S. S. Lee and J. F. Smith: *App. Physic Letters*, 1984, **44**, 296.
13. R. E. Green Jr: *Treatise on Materials Science and Technology Vol 3, Ultrasonic Investigation of Mechanical Properties*, Academic Press, New York, 1973, 73.
14. R. E. Green: *Novel Techniques of Non-destructive Examination*, E. A. Ash and C. B. Scruby, eds, Royal Society, London, 1986, 168 (Discussion).
15. M. E. Todaro, M. A. Doxbeck and G. P. Capsimalis: *Residual Stress in Design, Process and Materials Selection*, W. B. Young, ed., ASM Int., 1987, 59.
16. H. Barkhausen: *Phys. Zeitschrift*, 1919, **20**, 401.
17. D. J. Buttle, G. A. D. Briggs, J. P. Jakubovics, E. A. Little and C. B. Scruby: *Novel Techniques of Non-destructive Examination*, E. A. Ash and C. B. Scruby, eds, Royal Society, London, 1986, 363.
18. K. Tiitto: *Residual Stress in Design, Process and Materials Selection*, W. B. Young, ed., ASM Int., 1987, 27.

PREDICTION OF SURFACE QUALITY THROUGH CUTTING CONDITIONS IN THE GRINDING PROCESS

M. TAWFIK
Engineering Department, Institute of Productivity, Zagazig University, Egypt

ABSTRACT
In the grinding process, the resulting surface quality is attributed to the cutting conditions, parameters of the grinding wheel and properties of the work material. The surface roughness in surface grinding is predicted through application of the micromilling theory in case of single grit, and via wear theory as another approach.

This paper includes an experimental investigation of surface roughness at different values of depth of cut and hardness of work material. The surface roughness for the conditions of experimentation was determined applying both the micromilling and wear theories. The theoretical and experimental results have been analysed and compared, in order to evaluate the effectiveness of these prediction models.

1. INTRODUCTION

Grinding is applied for improvement of geometric accuracy and surface finish. Under proper conditions, an accuracy of 0.002 μm and surface roughness of 0.05 μm R_a (roughness average number) can be achieved.[1,2] As is well known, grinding is classified into cylindrical (external–internal) and flat surface grinding which can be developed either by the periphery of the grinding wheel or by its face.

The geometry of chip formation in grinding is similar to that of the milling process as the sharp edges of abrasive grits projecting from the grinding wheel surface act as extremely small cutting teeth. On the other hand, the cutting action of grinding differs from the milling action in the random geometry of the grits cutting angle, much smaller pitch of grits edges and smaller chip size. The grits are sufficiently hard to cut steels of hardness up to 850 HV. The surface roughness achieved in grinding depends mainly on the properties of the grinding wheel (grit size, hardness, kind of bond, wheel dressing and trueing, etc.), grinding conditions (cutting speed, depth of cut, feed, cooling conditions, etc.) the work material and technical state of the grinding machine.

2. THEORETICAL ANALYSIS

In surface grinding, two approaches may be used for theoretical prediction of surface roughness. The first is based on the theory of micromilling, while the second is derived referring to wear theory.

2.1 Prediction of Surface Roughness on the Basis of Micromilling Theory

In this approach, the surface roughness is predicted in terms of the maximum chip thickness. The assembly of grits in a grinding wheel of diameter D is considered to lie on the periphery with projecting points spaced $1/N$ mm apart, where N is the number of grits per mm of wheel circumference (Fig. 1).

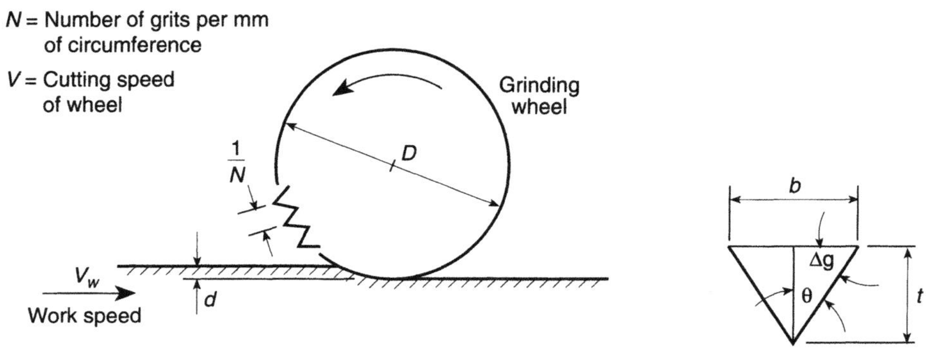

Fig. 1. Grits location and geometry.

By analogy with milling and simulating the grit to a single point tool, the maximum chip thickness t is determined as follows:[1]

$$t = 2f_g \sqrt{d/D}$$

where f_g is the feed of work per grit and d is the depth of cut. f_g is given by the following relationship:

$$f_g = \frac{V_w}{N\ V}$$

where V_w is the work speed (table speed) and V is the cutting speed, expressed as:

$$V = \frac{\pi D_n}{60 \times 1000}\ (\mathrm{ms}^{-1}) \text{ or } V = \frac{\pi D_n}{1000}\ (\mathrm{m\ min}^{-1})$$

where n is the rotation speed of grinding wheel, rpm and D in mm. Substituting the expression of f_g in the above equation of t, we get:

$$t = \frac{2V_w}{N\ V}\sqrt{\frac{d}{D}}$$

Therefore, the area of the metal removed across the grinding wheel of width B in place of the maximum chip thickness, A_r, would be determined as follows:

$$A_r = tB = \frac{2V_w\ B}{N\ V}\sqrt{\frac{d}{D}} \tag{1}$$

On the other hand, A_r can be determined regarding that it is achieved by a succession of grits in which the area of cut by a single grit at the maximum chip thickness is $tb/2$ where b is the width of cut area (Fig. 1). Introducing the average ratio (r) of width b to the depth of cut area t, this area will be expressed as $t^2\ r/2$ and consequently the total area of the metal removed (A_r) will be obtained as follows:

$$A_r = \frac{N\ B\ t^2 r}{2} \tag{2}$$

Equating (1) and (2) and solving for t, we get:

$$t = \left[\frac{4V_w}{V\ Cr}\sqrt{\frac{d}{D}}\right]^{1/2}$$

where C is the number of effective grits per mm^2 of grinding wheel surface: $C = N^2$. N is approximated by the surface peak count, Pc. The ratio r is determined by laboratory testing and C is measured experimentally by rolling the grinding wheel on a piece of smoked glass and counting (under the microscope) the marks left by the grit points.

The maximum depth of grooves cut in the work during surface grinding must be of the order of maximum chip thickness, or the peak-to-valley average reading of surface finish. Experience shows that the surface roughness average number, R_a, lies between 1/3 and 1/5 of the peak-to-valley value. Therefore, R_a would be predicted as:

$$R_a = \left(\frac{1}{3} - \frac{1}{5}\right) t = \frac{1}{4}\ t$$

or: (3)

$$R_a = \frac{1}{4}\left[\frac{4V_w}{V\ Cr}\sqrt{\frac{d}{D}}\right]^{1/2}$$

2.2 Prediction of Surface Roughness on the Basis of Wear Theory

In this case, the surface finish parameter R_a is determined as follows:[3]

$$R_a = \frac{K\ \sqrt{W/H_m}}{\tan\ \theta}\quad (\mu\text{m}) \tag{4}$$

where k is a constant, W is the normal load on surface, H_m is the hardness of the work material and θ is the half angle of grit tip (Fig. 1). K and θ are determined as:

$$k = \frac{D}{4.43},\ \theta = 90 - \Delta q \quad \Delta q\text{: average slope of surface}$$

The normal lead W is determined from: the volume rate of wear V_L (volume of removed material on surface per unit sliding distance), given by the following relationship:

$$V_L = \frac{0.5\alpha\ \beta\ W \cot\ \theta}{H_m} \tag{5}$$

where α is the ratio of the actual cutting force to the theoretical one and β is the scratching efficiency (ratio of the metal removed to the total groove volume). On the other hand, the volume rate of wear per unit time V_T can be expressed as follows:

$$V_T = A_c\ V_w$$

where A_c is the chip area given as:

$$A_c = S_0 d \quad \text{So: cross feed in mm/traverse}$$

Since the velocity of grinding wheel (speed of cut) V is much greater than the table speed V_w, then V_L can be obtained dividing V_T by V and hence:

$$V_L = \frac{S_0 d\ V_w}{V} \tag{6}$$

Equating (5) and (6) and solving for W we get:

$$W = \frac{2S_0 d\ H_m\ V_w}{\alpha\ \beta\ \cot\ \theta\ V} \tag{7}$$

3. EXPERIMENTAL WORK

The experimental programme was designed to provide the data necessary for verification of theoretical models of surface quality prediction. The experimentation comprised a set of measurements of the surface roughness parameter R_a at different values of depth of cut and hardness of work material, and at constant values of the other parameters included in the prediction models. It is to be noticed that hardness appears explicitly in the wear theory model and implicitly in the micromilling model.

3.1 Tested Specimens

Several specimens of constructional steel BS 080 M30 HR of ultimate strength 490 MPa and hardness 143 to 192 HB were prepared with dimensions of 10 cm length × 5 cm width × 1 cm thickness each. To achieve different values of hardness, the specimens were subjected to quenching followed by tempering at temperatures of 600, 550, 500, 450, 400, 350 and 300°C. Hardness of specimens was measured on the Brinell scale and its values corresponding to the mentioned tempering temperatures were found to be 269, 370, 410, 436, 487, 534 and 565 HB respectively.

3.2 Test Conditions

Grinding of specimens was carried out using the flat surface grinding machine type 3Г71 of horizontal spindle with grinding wheel type 250 × 25 × 75 A9946 K 9 V. The rotation speed of the wheel (n) was 2700 rpm. The work (table) speed (V) was maintained to 14 m/min. The cross feed (S_0) during tests was adjusted to 15 mm/traverse. The specimens of each group of specified hardness were ground at eight values of depth of cut of 0.01, 0.02, 0.035, 0.05, 0.075, 0.1, 0.125 and 0.15 mm.

3.3 Measurement of Surface Roughness

This was carried out using a surface finish tester type Talysurf. 6, manufactured by Rank Taylor Hobson. For each group of specimens of certain hardness, and at each of the test values of depth of cut, the parameters R_a, Δq, P_c were measured across the lay. Three successive measurements of each roughness parameter were conducted and the mean value was considered.

4. APPLICATION OF THEORETICAL MODELS

For assessment of the effectiveness of the theoretical models for surface quality prediction, the measured values of P_c and Δq together with the assigned values of grinding parameters D, n, V_w and S_0 were used to calculate R_a applying the micromilling theory model (equation (3)) and wear theory model (equation (4)) for the test values of depth of cut and material hardness. The normal load W in equation (4) was calculated by equation (7) taking unity for each of the coefficients α and β. A computer program for calculations was implemented.

5. RESULTS AND DISCUSSION

5.1 Investigation of Theoretical Models

For the purpose of comparison, the experimental results together with the theoretically calculated results according to the two prediction models (micromilling – wear) were interpreted on common graphs in the form of variation of the roughness parameter R_a with the depth of cut d for the different values of hardness. This is depicted in Figs 2 to 8.

Concerning the correlation between experimentation and theoretical prediction, the figures reveal clearly much better agreement for results achieved by the micromilling model with experimental results, compared with the wear model. This can be accounted for by deviations in the concepts for determination of wear rate suggested by the wear theory from the actual conditions of wear action in the grinding process. The results of calculations by the micromilling theory coincide with the experimental results in the law of variation of R_a, but with smaller values. Higher deviation is found at smaller depths of cut. Larger experimental values compared with theoretical results are attributed to the fact that the real factors participating in the grinding process such as the plastic deformation of material and the actual geometry and distribution of abrasive grits cannot be simulated completely in the theoretical model.

For quantitative evaluation of the increase in the actual values of R_a, obtained by experiments (Raex) over the theoretically calculated results by micromilling theory (Rath), the ratios Raex/Rath have been calculated for the different values of depth of cut d and material hardness. Figures 9, 10 and 11 illustrate the variation of Raex/

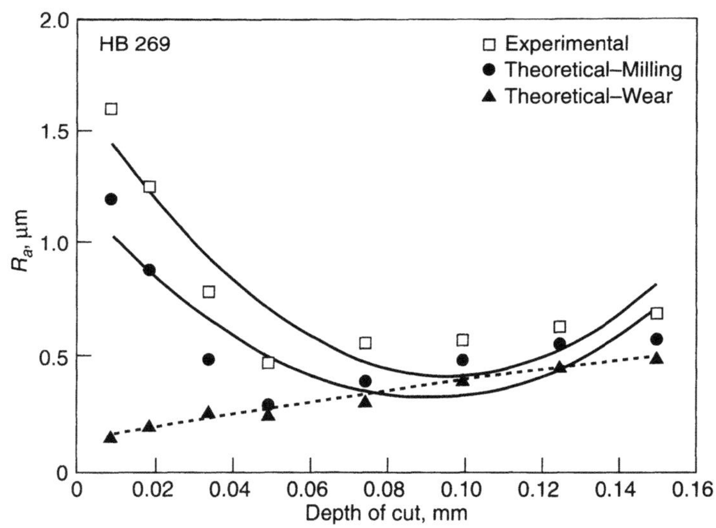

Fig. 2. Variation of experimental and theoretical surface roughness parameter R_a, hardness 269 HB.

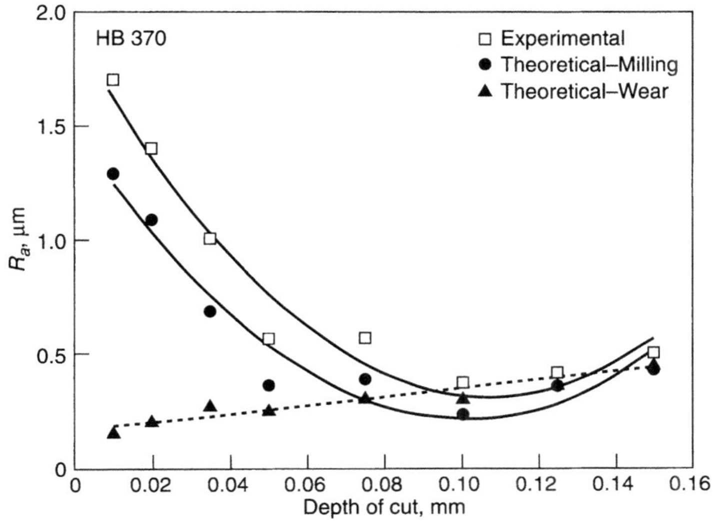

Fig. 3. Variation of experimental and theoretical surface roughness parameter R_a, hardness 370 HB.

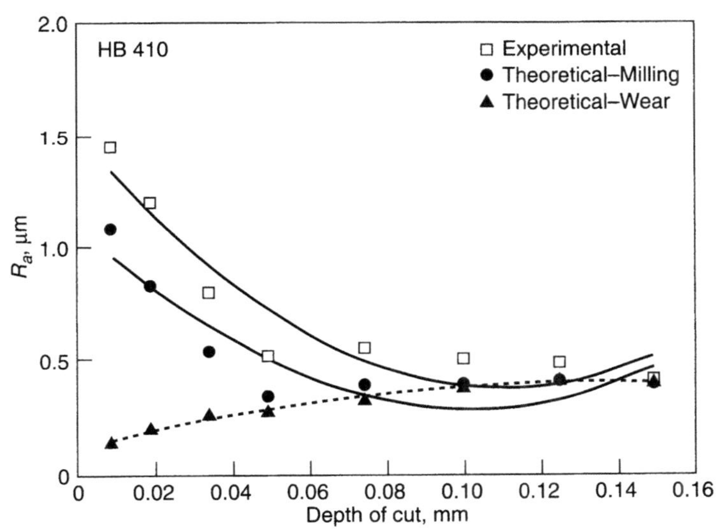

Fig. 4. Variation of experimental and theoretical surface roughness parameter R_a, hardness 410 HB.

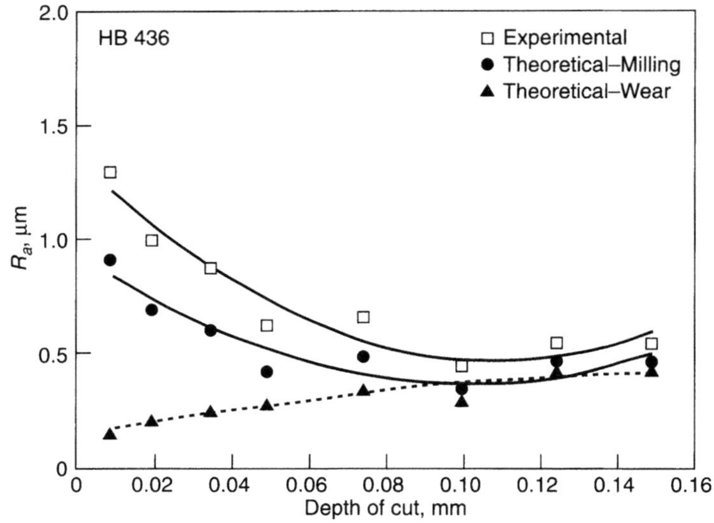

Fig. 5. Variation of experimental and theoretical surface roughness parameter R_a, hardness 436 HB.

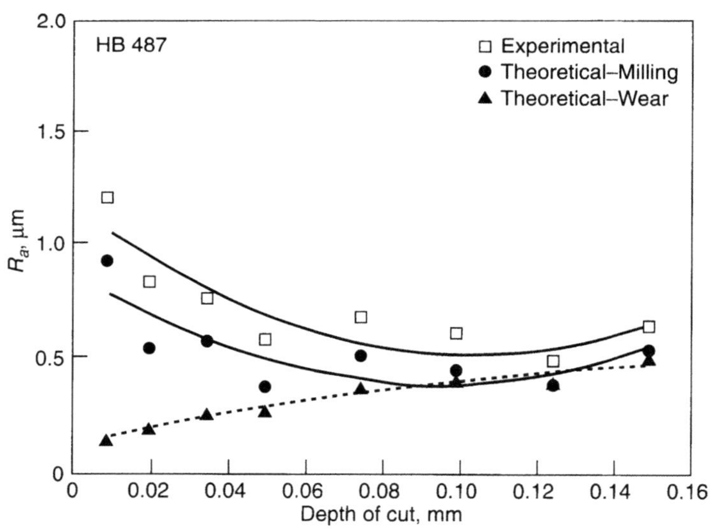

Fig. 6. Variation of experimental and theoretical surface roughness parameter R_a, hardness 487 HB.

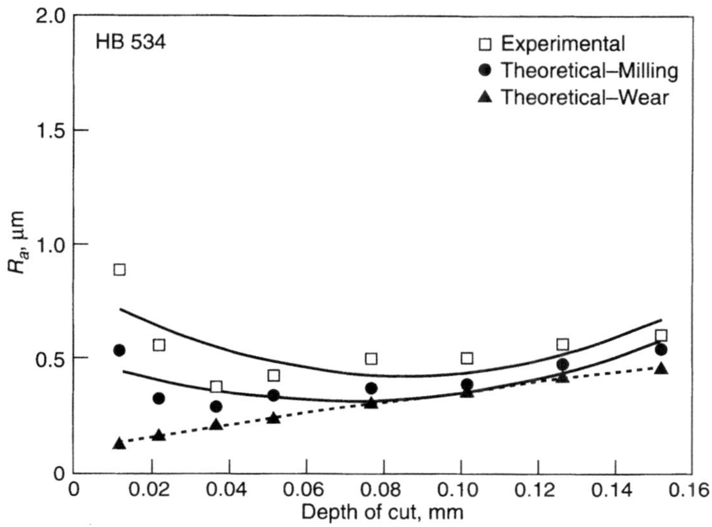

Fig. 7. Variation of experimental and theoretical surface roughness parameter R_a, hardness 534 HB.

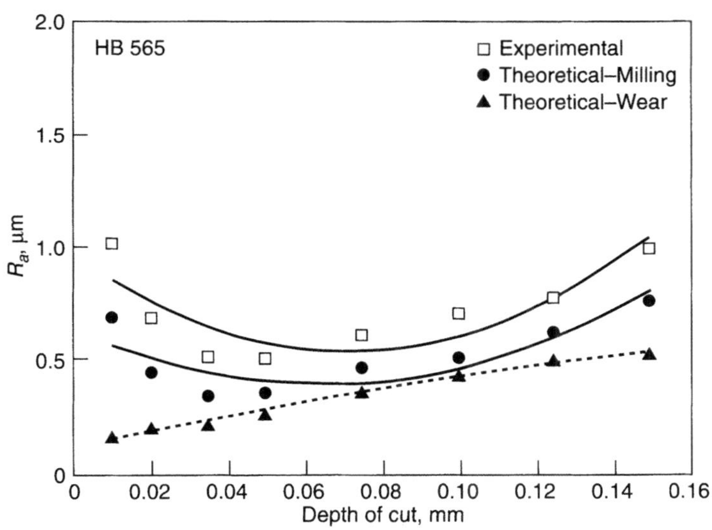

Fig. 8. Variation of experimental and theoretical surface roughness parameter R_a, hardness 565 HB.

Rath with *d* for the different values of hardness, while Figs 12, 13 and 14 indicate the variation of this ratio with the hardness (after rearrangement of results) for the different values of depth of cut.

Figures 9, 10 and 11 show quasi-linear decrease of the ratio Raex/Rath with the depth of cut *d* for all values of hardness. This ratio varies from 1.16 (at higher values

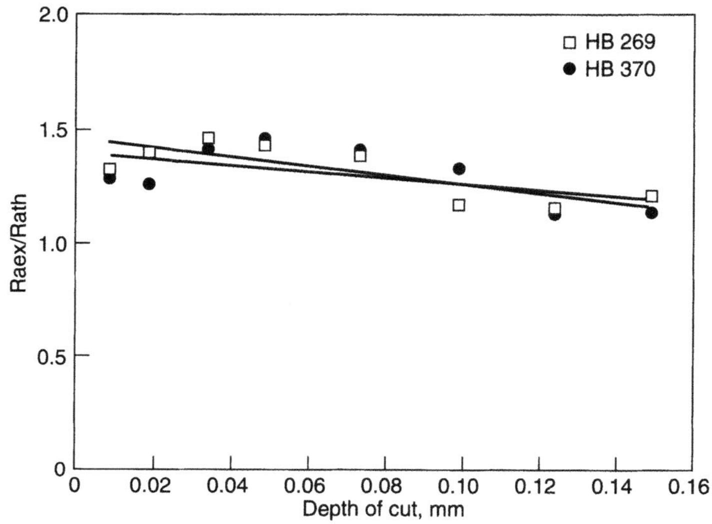

Fig. 9. Variation of Raex/Reth with depth of cut, hardness 269, 370 HB.

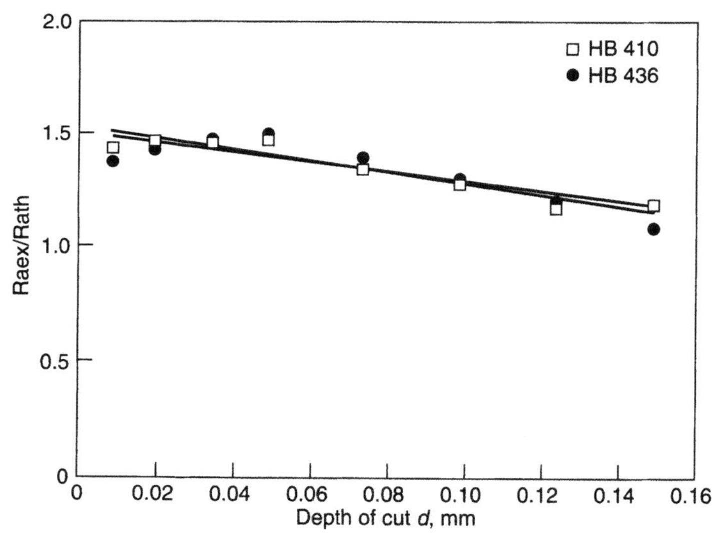

Fig. 10. Variation of Raex/Reth with depth of cut, hardness 410, 436 HB.

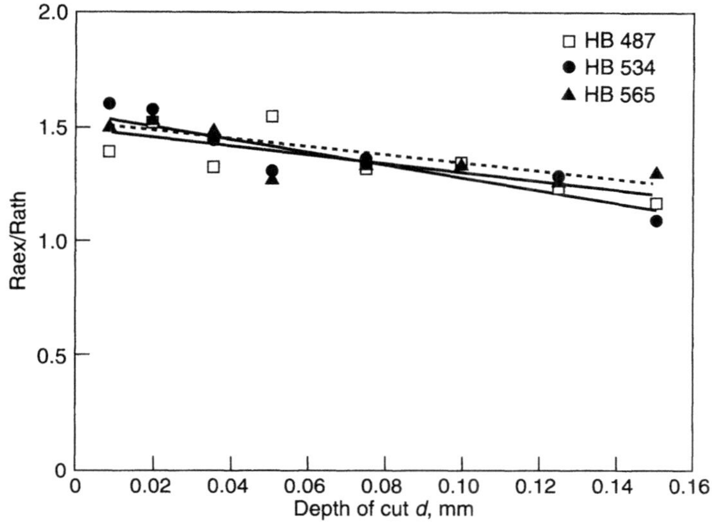

Fig. 11. Variation of Raex/Reth with depth of cut, hardness 487, 534, 565 HB.

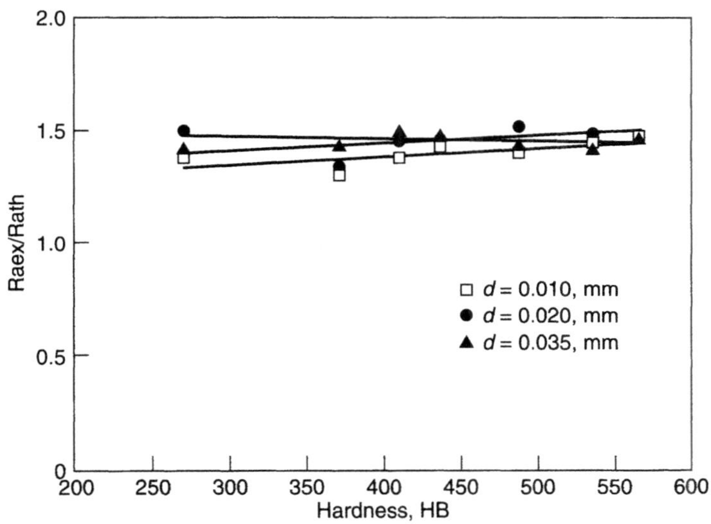

Fig. 12. Variation of Raex/Reth with hardness, d = 0.01, 0.02, 0.035 mm.

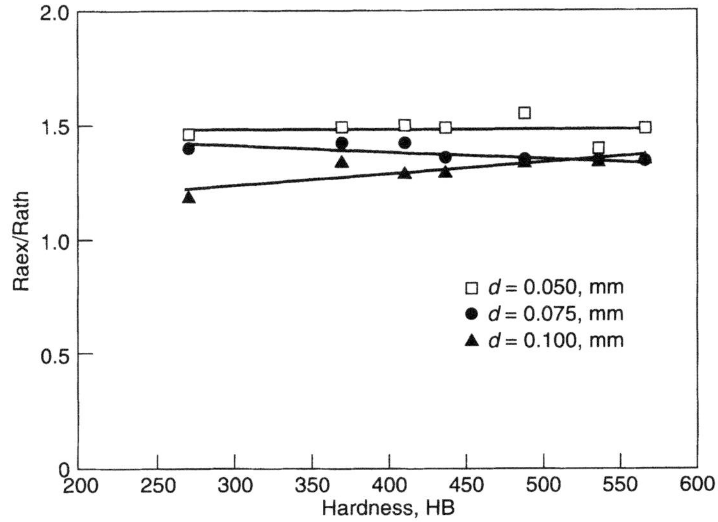

Fig. 13. Variation of Raex/Reth with hardness, d = 0.05, 0.075, 0.1 mm.

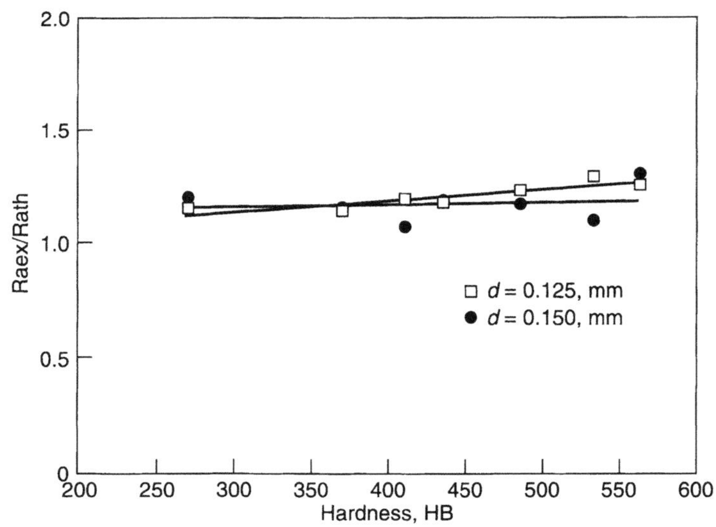

Fig. 14. Variation of Raex/Reth with hardness, d = 0.125, 0.15 mm.

of *d*) to 1.49 (at smaller *d*). The variation of Raex/Rath with *d* is found to follow the relation

$$\frac{\text{Raex}}{\text{Rath}} = 1.5 - 2.1\,d$$

Figures 12, 13 and 14 show slight variation of the ratio Raex/Rath with hardness (within 5% at maximum) and hence, in view of the simplifications, the ratio can be considered to be independent of hardness.

5.2 Modification of Micromilling Model

Because of the larger actually measured values of surface roughness parameter relative to the results calculated by the micromilling formula, a correction factor *u* is proposed to be introduced in this formula, which would take the form:

$$R_a = \frac{u}{4}\left[\frac{4\,V_w}{V\;Cr}\sqrt{\frac{d}{D}}\right]^{1/2}$$

The correction factor *u* is always higher than unity and depends mainly on the depth value of cut *d* with a variation identical to that of the ratio Raex/Rath presented above. According to the results achieved in this work the value of *u* lies between 1.16 and 1.49.

6. CONCLUSIONS

1. An analysis of the theoretical models for prediction of surface quality in terms of the roughness average number according to the micromilling and wear theories, has been conducted.
2. An experimental programme was carried out to provide the essential data for application of theoretical models in order to investigate their effectiveness and applicability. The tests comprised measurements of average surface roughness numbers at different values of depth of cut and material hardness.
3. The surface roughness average number was calculated using both the micromilling and wear models.
4. Experimental results have been compared with theoretical values achieved by these two models, when the results of application of the micromilling model were found to agree functionally with experimental results but with somewhat reduced values. The wear theory model yielded significant discrepancy from practical results. Accordingly, the micromilling model was considered more appropriate for theoretical prediction of surface finish.
5. The larger values of experimental results over the theoretical values predicted via the micromilling model necessitate modification of this model, which has been

developed through introducing a correction factor greater than unity into the model formula. This correction factor depends mainly on the depth of cut and varies from 1.16 to 1.49 for the range of parameters used in this work.

REFERENCES

1. A. J. Lissaman and S. J. Martin: *Principles of Engineering Production*, Hodder and Stoughton, London, 1982.
2. V. V. Loskutov: *Grinding of Metals*, Mir Publishers, Moscow, 1981.
3. T. C. Buttery *et al.*: 'Some Effects of Deterioration in Abrasive Surfaces on Cutting Efficiency and Surface Finish', *Proc. 18th MTDR Conf.*, September 1977, London, 439.
4. J. Peklenik: 'The Statistical Mechanism of Chip Formation in Grinding', *Proc. Int. Conf. Prod. Eng.*, Tokyo, 1974.
5. T. C. Buttery and J. F. Archard: 'Some Microscopical Investigations of Grinding and Abrasive Wear', *Journal of Microscopy*, 1971, **94**(1), 13.
6. L. E. Samuels: 'Abrasive Surface Finishing Processes: Mechanism', *Proc. Conf., Prod. Sci. In Industry*, Monash University, Australia, 1971, 247.
7. K. Nakayama and M. C. Shaw: 'Study of the Finish Produced in Surface Grinding. Pt. 2', *Proc. Inst. Mech. Eng.*, 1967–68, **182**(3K), 179.

QUALITY CONTROL AND QUALITY ASSURANCE IN HEAT TREATMENT

GRAHAM W. CONRAN
Engineering Technology Advisor, Royal Australian Mint, Canberra, Australia

ABSTRACT

Heat treatment has long been known as the 'Black Art', no doubt receiving such a sobriquet from the earthen floor, deeply ingrained with oil, coal dust and cast iron chips and the dingy, fume-laden atmosphere of the heat treatment shop of the 19th and early 20th centuries. The 'art' of the heat treater has given way firstly to the craftsman and now to the technologist as newer, better and cleaner processes have evolved.

In many cases heat treatment has been recognised as a 'technology' in its own right and a modern heat treatment facility is clean and bright with good housekeeping and free from atmospheric pollution. Although equipment calibration and control is now widely practised, few heat treaters are prepared to guarantee their work either due to lack of technical knowledge or by not fully understanding the capabilities of their equipment.

This paper looks at the requirements of AS 3902, particularly with respect to captive heat treatment facilities, and to examine what the modern heat treater must be prepared for as quality control and quality assurance requirements become more stringent. As heat treatment has become a technology so the heat treater must become technologically aware and be prepared to offer far more support, help and advice to his customers resulting in the long run to guarantees being offered.

1. INTRODUCTION

In the current economic climate where competition is forcing economists and accountants to control the financial viability of their companies more closely than ever before, a 1% decrease in scrap generation can often mean the difference between a borderline operation and a healthy profit margin. Too often in the past, it was hoped that every batch of heat-treated work would be satisfactory, but always making an allowance for the odd batch failing. The failure of a die or critical machine component in heat treatment was often greeted with a shrug of the shoulders, a few choice remarks and little else.

The need to change this perception of our customers, whether they are from within our own company, as is the case with captive heat treatment shops, or are commercial customers, and the attitude of the heat treater is now essential if heat treatment is to be a credible technology in the future. What has happened to make the need for change so imperative?

The following five points are made in an attempt to answer this question.

1. The need to increase productivity is well known and the reduction in the generation of scrap in heat treatment will achieve this.
2. Together with the need to increase productivity is the need to reduce costs and hence increase profits and a greater efficiency in heat treatment operations will achieve this.
3. New and better equipment is now available and our expertise must keep pace with such improvements. For instance, when salt baths were the only way to heat treat high-speed steel only single components were likely to fail due to faulty heat treatment, whereas now we are hardening high-speed steel in batches of 100 kg or more with consequent dire results if the treatment is faulty.
4. The wide, and in time universal, acceptance of AS 3902 must be taken into account. Within a short time a lack of accreditation to this standard will see companies with in-house heat-treatment facilities being forced to the wall due to the lack of major customers and the commercial heat treater existing on small jobbing orders where quality is not a prime consideration.
5. As new metals and alloys are becoming more available for general use the heat treater must keep abreast of the metallurgical aspects of these if he is to understand the heat treatment processes required and to be able to adjust to changes in conditions. Failure to do this, particularly in the case of the commercial heat treater, will inevitably lead to greatly reduced orders and possible bankruptcy.

Those of us who have worked in the heat-treatment industry have often heard the saying that '90% of tooling failures are due to heat treatment'. This is not to say that these failures are all due to bad heat treatment as design, wrong material, prior treatment (or lack thereof), bad machining practice and bad material selection can all cause failure.

However, it is essential for the modern heat treater to be aware of these problems and be able to advise his customers accordingly, even to the point of refusing work if his reputation could be jeopardised.

Let us look at these five points in more detail.

1.1 Increase in Productivity

Productivity can be increased in a number of ways; increasing the number per load, changing the cycle, changing the temperature (as in carburising) and devising better racking and loading systems being some methods. In increasing the number per load one must be aware of the heating capacity of the furnace, the ability of the load to heat and quench evenly, the ability of any atmosphere to circulate evenly through the charge and the specification requirements with respect to the spread of hardness or case depth allowed.

The limited ruling section for through-hardening must be known for certain quenching rates and whether it is possible to use a faster quenching medium to accommodate the larger load without causing cracking or distortion. In the case of carburising one must be aware of the upper temperature limits and the probability of grain growth and whether the components can be directly quenched or must be grain refined and homogenised.

To attempt changes in productivity it is clear that the heat treater must have a very thorough knowledge of his equipment as well as a sound knowledge of the metallurgy of heat treatment. The reduction of rejects by the correct application of this knowledge can be often more productive than changes in processes or increasing load size.

1.2 Increase in Efficiency

In this case increases in efficiency can be obtained by better planning of work, optimising loads, utilising the dark hours by using programmable controllers, ensuring optimum atmosphere flows and using time clocks to switch furnaces on and off.

An increase in efficiency coupled with a productivity increase reduces costs and can usually be achieved at little or no capital cost.

1.3 New and Better Equipment

Since World War 2 the changes to heat treatment practice have been dramatic. Gone are the dark, dingy, smoky, hot and dirty heat treatment shops. Gone are the days when the heat treater was at best a qualified blacksmith. Fortunately we now have well laid-out shops with good ventilation, clean, dust-free conditions and in particular far superior furnace and testing equipment. We now have vacuum furnaces, fluidised bed furnaces, the TD process, the NITROTEC process, CVD, PVD, Plasma Ion Nitriding, Ion Implantation and gaseous nitriding to name some of the processes with which the heat treater has to contend.

While all of these processes are never likely to be available in the one facility the heat treater should have a knowledge of them so that he can advise his management (in the case of the in-house facility) or his customers of what they can offer to either increase productivity or reduce costs by the utilisation of different materials.

Better designs and construction, more efficient insulation, better control equipment, more accurate and stable atmosphere controllers and the use of digital read-out instruments all combine to allow furnaces to have much better temperature uniformity, less cyclic temperature variation and better spatial and temporal temperature variations than ever before.

This allows the heat treater to have tight control over his processes as long as checking and calibration of the equipment is carried out in some regularised manner. The better the equipment the greater the need for control, 'Quality Control'.

The control of quality does not end with the best furnaces and controls but also with the test equipment such as hardness testers, impact testers, tensile testers and so on. These must also be checked and calibrated and the results obtained from them interpreted correctly.

Inevitably we come back to the point that no matter how good our furnaces and test equipment may be they cannot guarantee 'Quality' and that only with the expertise and regard for 'Quality' can 'Quality Control' in heat treatment be achieved.

1.4 AS 3902

This standard is identical with ISO 9002 and is titled: 'Quality Systems for Production and Installation'. Already in Australia orders have been lost because accreditation to the standard has not either been achieved or applied for. This will be the norm in the very near future and without accreditation no worthwhile orders will be obtained.

The standard, in Clause 1.2, states:

'This International Standard is applicable in contractural situations when –
(a) the specified requirements for product are stated in terms of an established design or specification;
(b) confidence in product conformance can be attained by adequate demonstration of a certain supplier's capabilities in production and installation.'

The standard clearly places heavy emphasis on the audit trail and the need for 'Work Instructions' and a 'Quality Manual' are obligatory. As a matter of fact, the Royal Australian Mint has prepared data sheets as part of their accreditation requirements to ensure such practices are indeed carried out.

The total documentation for AS 3902 is too long to detail here but in the context of heat treatment Clause 4.10 is reproduced.

4.10 Inspection, Measuring and Test Equipment

The supplier shall control, calibrate and maintain inspection measuring and test equipment, whether owned by the supplier, on loan, or provided by the purchaser, to demonstrate the conformance of product to the specified requirements. Equipment shall be used in a manner which ensures that measurement uncertainty is known and is consistent with the required measurement capability.

The supplier shall –
(a) identify the measurements to be made, the accuracy required and select the appropriate inspection, measuring and test equipment;
(b) identify, calibrate and adjust all inspection, measuring and test equipment and devices that can affect product quality at prescribed intervals, or prior to use, against certified equipment having a known valid relationship to nationally recognised standards – where no such standards exist, the basis used for calibration shall be documented;
(c) establish, document and maintain calibration procedures, including details of equipment type, identification number, location, frequency of checks, check method, acceptance criteria and the action to be taken when results are unsatisfactory;

(d) ensure that the inspection, measuring and test equipment is capable of the accuracy and precision necessary;
(e) identify inspection, measuring and test equipment with a suitable indicator or approved identification record to show the calibration status;
(f) maintain calibration records for inspection, measuring and test equipment (see 4.15);
(g) assess and document the validity of previous inspection and test results when inspection, measuring and test equipment is found to be out of calibration;
(h) ensure that the environmental conditions are suitable for the calibrations, inspections, measurements and tests being carried out;
(i) ensure that the handling, preservation and storage of inspection, measuring and test equipment is such that the accuracy and fitness for use is maintained;
(j) safeguard inspection, measuring and test facilities, including both test hardware and test software, from adjustments which would invalidate the calibration setting.
Where test hardware (e.g. jigs, fixtures, templates, patterns) or test software is used as suitable forms of inspection, they shall be checked to prove that they are capable of verifying the acceptability of product prior to release for use during production and installation and shall be rechecked at prescribed intervals. The supplier shall establish the extent and frequency of such checks and shall maintain records as evidence of control (see 4.15). Measurement design data shall be made available, when required by the purchaser or his representative, for verification that it is functionally adequate.

As you can see, this Clause covers inspection, measuring and test equipment and in practice all of our heat-treatment operations revolve around them. We measure temperature and hardness, we use test equipment such as hardness testers, tensile testers and impact testers and we inspect the product of our heat treatment for hardness, straightness, case depth and so on. In other words, the standard applies to heat treatment processes as much as to any other processes.

The first sentence of this Clause, 'The supplier shall control, calibrate and maintain . . .' is the heart of the matter and every heat treatment facility must have the capability and procedures in place to meet the requirements as laid down.

1.5 New Metals and Alloys

As the newer metals and alloys become more widely available the need for the heat treater to keep adding to his metallurgical knowledge becomes essential if he is to be capable of offering a quality service to his customers.

The term 'Heat Treatment' has a very wide scope and includes any operation involving change of temperature with the object of producing an alteration in physical properties.

This definition opens the whole field of surface engineering to the heat treater with all its many fields and branches. Certainly it is not expected that every heat treater must become proficient in every field but it is becoming more and more essential for him to at least have a basic understanding of a number of the better known processes and to have a reasonable grasp of the metallurgy of iron and steel.

Clause 4.17 of AS 3902 states:
'The supplier shall establish and maintain procedures for identifying the training needs and provide for the training of all personnel activities affecting quality during production and installation. Personnel performing specific assigned tasks shall be qualified on the basis of appropriate education, training and/or experience, as required. Appropriate records of training shall be maintained (see 4.15)'.

2. CONCLUSION

'Quality Control' in the heat-treatment field is not now something that we give lip service to, but a fact of life. With all the advances in other fields requiring that quality must not only be maintained but enhanced we in the heat-treatment field cannot any longer put our heads in the sand and expect the current status quo to be maintained into the future. The heat treatment industry in Australia is on the whole held in very poor esteem by its customers and if we are to change that image we must be prepared to fund new equipment, up-grade existing equipment, train staff and to finally obtain accreditation to AS 3902. With accreditation comes the necessity to guarantee our work.

How many heat treaters do you know who at this point in time are prepared to guarantee their work?

If we are to be able to compete in any marketplace in the future our heat treatment facilities must be accredited to AS 3902 which means that 'confidence in product conformance can be attained by adequate demonstration of a certain supplier's capabilities in production and installation.'

If one were to coin a phrase for the heat treatment industry it might well be 'Spend to survive'.

QUALITY ISSUES WITH THE PLASMA NITRIDING PROCESS

NEE LAM LOH and BAHN JEE TAN
Nanyang Technological University, School of Mechanical & Production Engineering, Singapore

DANIEL NG
Doxon Engineering Pte Ltd, Singapore

ABSTRACT

Appropriate quality issues for advanced manufacturing processes and products are an open topic. In general the world of technology sees it necessary to establish some acceptable levels of norms and practices which should put all participating industries on a comfortable level of assurances and confidence in terms of quality and reliability. Hence, acquisition of the ISO 9000 standard is very much sought after; the adherence to which renders the participating company a certain and definite level of competence and provides, for the buyer, a certain level of confidence.

This paper, however, looks into some technical aspects of quality control insofar as the plasma nitriding process is concerned. Whilst each client would express different levels of expectation in terms of accuracy and consistency, it is concluded that some degree of prudence and professionalism would go a long way in enhancing the acceptability of the process.

1. INTRODUCTION

The heat treatment of engineering materials has been and often continues to be treated industrially as a kind of inexact science. Hence, quality control is usually quite relaxed and taken with much sympathy and tolerance. With mass production of small parts, this may well be legitimate and it is not unusual to accept even up to a 25% rejection rate with, say, conventional gas carburising. Advanced technology which improves the accuracy, controllability and consistency of processing parameters, coupled with increasing knowledge on the subject, have in recent times enabled companies involved in heat treatment to expect a higher level of quality assurance.

Perhaps it is fair to say that the quality of a company's service or product marks it out to be inferior or superior to rival companies. Indeed, it is quality that will decide if a product or service is up-market or down-market. As this paper involves plasma nitriding and Doxon is the only company known to the authors which provides such services in Singapore, its approach to quality control shall be used for discussion. Indeed, it was and is the tight quality control so specified and practised which renders Doxon as a heat treatment company of high precision.[1] It must be emphasised though, that the material and points covered here are necessarily somewhat general and require appreciation of the plasma nitriding process. In this way, proprietary information has not been revealed and the authors would like to seek the readers' understanding, and apologise for any lack of particular detail.

The various quality-related issues raised during pre-treatment, treatment proper and post-treatment inspection are reviewed.

2. PRE-TREATMENT CHECKS AND PREPARATION

Before a plasma nitriding treatment, it is essential that all relevant information must be accurately ascertained – the material and its chemical composition, its thermal and service history, if any, tempering temperature, shape distortion tolerances, case depth required and any important surface finish to protect.

It must be emphasised that different materials respond differently to thermal treatment. Hence, it would be unacceptable, for example, to treat stainless steels together with a 1% Cr–Mo steel. Then again, it must be decided if the treatment is meant for fatigue, wear and/or corrosion resistances.

Knowledge of prior thermal and service histories are essential as this will determine if parts can be treated in a single load. Where the retention of core hardness is important, then the tempering temperature must be known. As a rule of thumb, the plasma nitriding temperature should be about 30°C lower than the tempering temperature. This is particularly important where the core hardness is required, not only to provide a good support for the case, but also to bear higher service loads. Under such a situation, the parts are required to be sufficiently hardened and tempered to ensure microstructural stability.

Where the parts have been serviced before, as with most dies and moulds undergoing trial runs, then it is unlikely that the surfaces would be clean. Extreme care must therefore be observed to remove all traces of hydrocarbons, foreign elements and grease, to avoid excessive arcing, which may bring about surface damages.

Case depths and surface hardnesses must be agreed upon. In this matter, there is always the tendency to over-specify and allow for subsequent grinding off of the parts to shape. In particular, both parties must agree on the definition of effective versus actual case depth.

2.1 Dimensional Tolerance

Plasma heat treatment offers good dimensional control. This is brought about by several factors. With plasma nitriding and nitrocarburising, these are ferritic thermochemical processes which do not involve any phase transformation. In addition, both the heating-up and cooling-down periods can be accurately controlled. Dimensional stability of the workpiece is further enhanced by monitoring the sputtering rate to balance the inevitable growth consequential on the impregnation by the medium. For these reasons, final growth per surface has been successfully reduced to less than 5 μm, with case depth reaching down to 100 μm or so.

Perhaps surprisingly, such good control in heat-treatment services is not too common in industrial practice. In one particular case, as shown in Fig. 1, 300 pieces

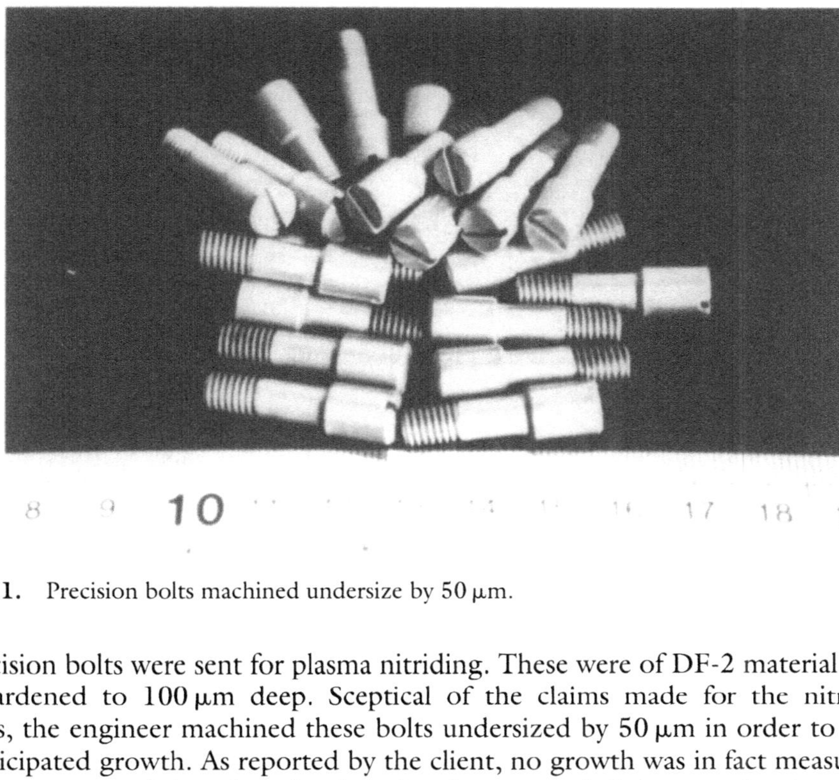

Fig. 1. Precision bolts machined undersize by 50 μm.

of precision bolts were sent for plasma nitriding. These were of DF-2 material to be case-hardened to 100 μm deep. Sceptical of the claims made for the nitriding process, the engineer machined these bolts undersized by 50 μm in order to allow for anticipated growth. As reported by the client, no growth was in fact measurable by micrometer; and the whole batch had to be scrapped. Subsequently, these were dimensioned to exact size for plasma nitriding.

2.2 Stress Relief

Where shape distortion is crucial and must be kept to the very minimum, then it is essential to insist on stress relief of the parts before plasma nitriding. After stress relief, it is necessary for the client is to inspect all dimensions to their satisfaction before plasma nitriding. 10–20 μm per surface growth must be allowed for; though growth down to 3–5 μm has been achieved with the plasma nitriding process.

Figure 2 shows some plasma nitrided groove cams made from JIS S45C carbon steel. The cams measured 140 mm in diameter and 9.0 mm in thickness. These were

Fig. 2. Groove cams plasma nitrided to tight dimensional control (courtesy of Doxon Engineering Pte Ltd).

stress relieved before final touch-up to precise dimension. Eight critical points were selected for measurement by CMM before and after plasma nitriding. An average growth of 8 μm was detected along the radius.

2.3 Surface Finish

There are occasions when surface finish is crucial. Treated parts normally show a uniform, matt feature. The final outcome of the surface topography is a function of the ionic species, the power input, the treatment time, the material and its hardness. Figure 3 is a schematic of the ionic interaction with cathodic surface; while Fig. 4 shows the sputtering yield of pure iron bombarded by N^+ and N_2^+ ions as a function of the ion energies at normal incidence.

Where a mirror polish is required, however, then a final lap after plasma nitriding would be necessary. In one particular instance, a mirror-polished surface was plasma nitrided, and the processing parameters were monitored to minimise the sputtering effect. Interestingly, the ionitrided surface was acceptable and the workpiece was put into service without any touch-up. Industrial experience also indicates that a final polish on a plasma nitrided surface normally takes only half the time to accomplish as compared to untreated parts. Figure 5 shows a mirror-polished part and the matt surface after plasma nitriding.

As the surfaces of the workpiece are continuously bombarded during the whole of the plasma heat treatment process, it is inevitable that surface roughening will occur.

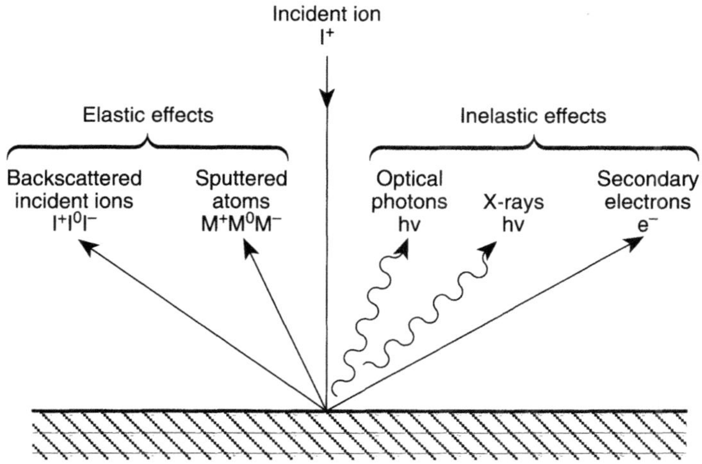

Fig. 3. Schematic of the ionic interaction with a cathodic surface.

This is particularly obvious as treatment temperature and time are increased. Figure 6 shows the scanning electron topography of a 1% Cr–Mo steel, after being bombarded in pure Argon for 20 h at 600°C, while Fig. 7 shows the surface topography of Inconel 718 plasma nitrided in nitrogen–hydrogen gas mixture for 16 h at 600°C.

Roughness is usually measured using the Form Talysurf Machine. It is normally expressed in terms of R_a, the arithmetic mean roughness value and R_t, the maximum peak-to-value height. Of the two, however, R_a is more commonly used industrially.

Experiments were carried out at the Nanyang Technological University to

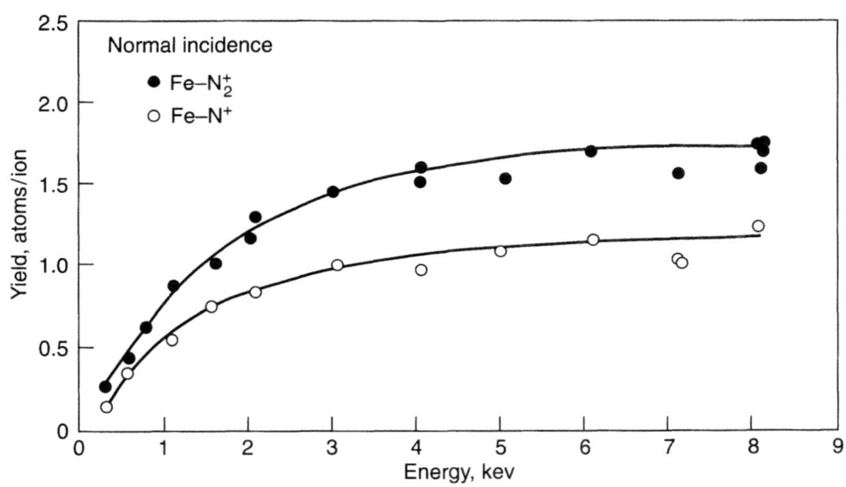

Fig. 4. Sputtering yield of iron bombarded by N^+ and N_2^+ ions as a function of ion energy at normal incidence after Bader *et al.*[2]

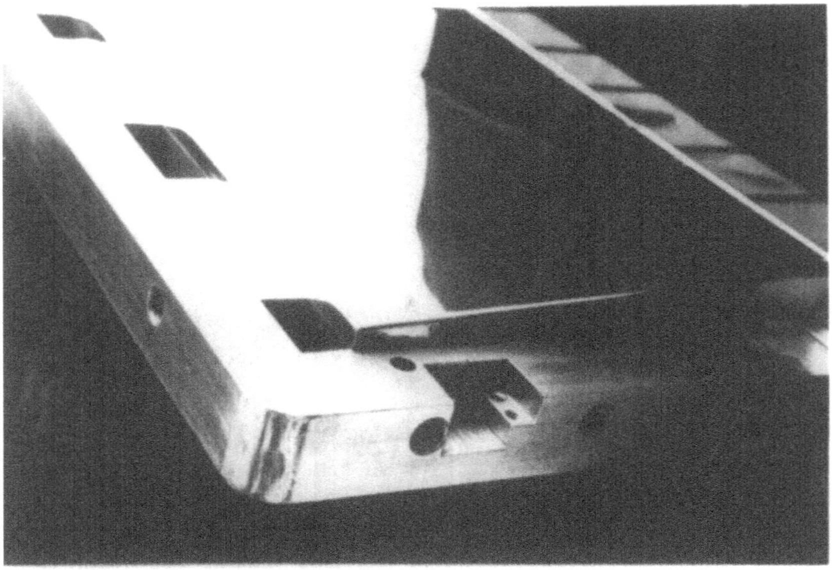

(a)

(b)

Fig. 5. A mirror-polished surface, (a) before and (b) after plasma nitriding.

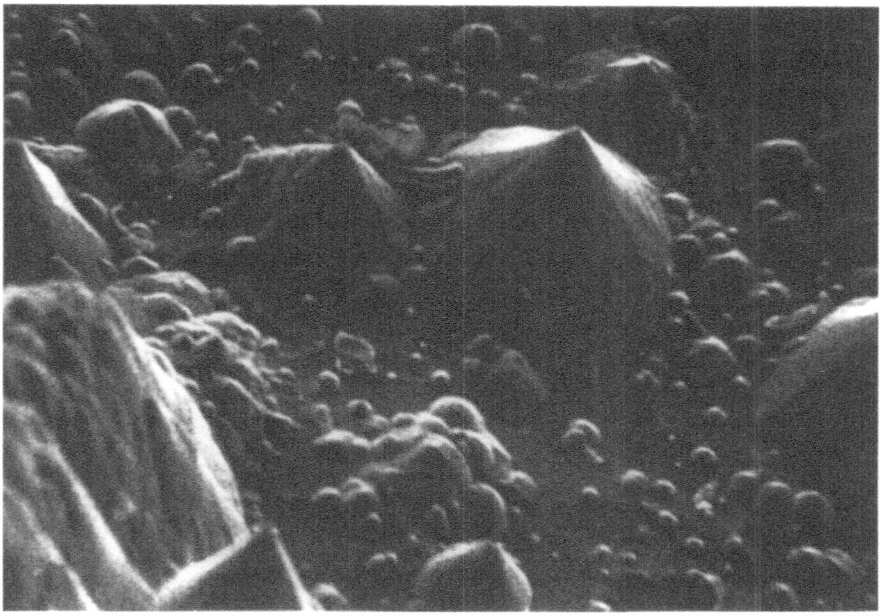

Fig. 6. Scanning electron micrograph of 1% Cr–Mo steel bombarded in pure Argon at 600°C for 20 h, 2.4 KVA (×500).

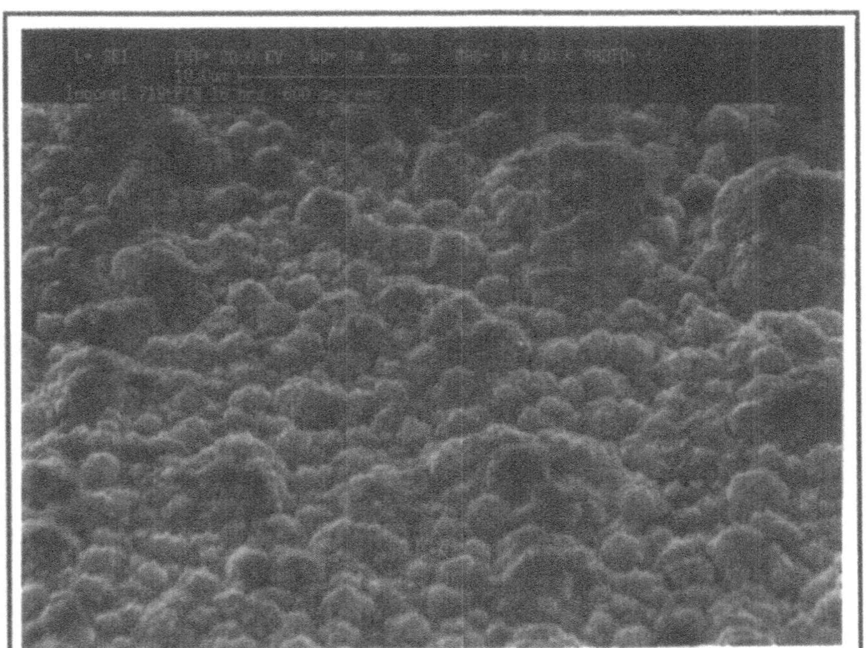

Fig. 7. Scanning electron micrograph of plasma nitrided Inconel 718.

evaluate the roughness effects of the various parameters such as power, gas composition and treatment time and temperature. Table 1 shows the chemical compositions of the 410 stainless steel and duplex stainless steel. The former was 280 VPN in hardness finished to an R_a value of 0.412, while the latter was only 190 VPN with $R_a = 0.175$. Tables 2 and 3 show the respective data collated with the corresponding parametric values. In all instances, no prior sputtering process was carried out and the gas composition and pressure were maintained constant. Four R_a values were measured to determine the respective average values.

Table 1. Chemical composition of the 410 stainless steel and the duplex stainless steel.

Element	C	Si	Mn	S	P	Cr	Ni	Mo	Ti	N
AISI 410 stainless steel (%)	0.12	0.31	0.66	0.004	0.017	11.79	0.6	0.03	–	–
Duplex stainless steel (%)	0.017	0.5	1.16	0.01	0.02	11.37	0.14	–	0.004	0.013

Table 2. Roughness values of the plasma nitrided 410 stainless steel.

S/no.	Temp (°C)	R1 (R_a)	R2 (R_a)	R3 (R_a)	R4 (R_a)	Rav (R_a)	Voltage (V)	Current (A)	Power (VA)
S0	400	0.676	1.604	0.518	0.341	0.785	200	1	200
S1	450	0.585	0.659	0.457	0.485	0.547	200	1	200
S2	480	0.363	0.567	0.746	0.557	0.558	200	1	200
S3	510	0.533	0.675	1.546	0.636	0.848	240	1	240
S4	540	0.866	0.782	0.418	1.006	0.768	240	1	240
S5	570	0.509	0.745	0.454	0.569	0.569	285	1	285
S12	580	0.589	0.491	0.745	0.616	0.61	285	1	285
S6	600	0.502	0.618	1.584	0.558	0.816	340	1.25	425
S13	630	0.596	0.657	1.75	0.662	0.916	300	1.1	330
S17	700	1.277	1.82	0.874	0.809	1.195	380–400	1.25	475–500

AISI 410 stainless steel
Initial roughness, $R_a = 0.412$

Table 3. Roughness values of the plasma nitrided duplex stainless steel.

S/no.	Temp (°C)	R1 (R_a)	R2 (R_a)	R3 (R_a)	R4 (R_a)	Rav (R_a)	Voltage (V)	Current (A)	Power (VA)
D0	400	0.237	0.227	0.379	0.277	0.28	200	1	200
D1	450	0.295	0.334	0.283	0.293	0.301	200	1	200
D2	480	0.328	0.314	0.203	0.282	0.282	200	1	200
D3	510	0.258	0.369	0.25	0.249	0.282	240	1	240
D4	540	0.336	0.224	0.574	0.33	0.366	240	1	240
D5	570	0.344	0.282	0.474	0.367	0.367	285	1	285
D12	580	0.332	0.381	0.412	0.299	0.356	285	1	285
D6	600	0.88	0.441	1.515	0.474	0.5775	340	1.25	425
D13	630	0.433	0.47	1.502	0.385	0.448	300	1.1	330
D17	700	0.667	0.945	0.751	0.694	0.764	380–400	1.25	475–500

Duplex stainless steel
Initial roughness, $R_a = 0.175$

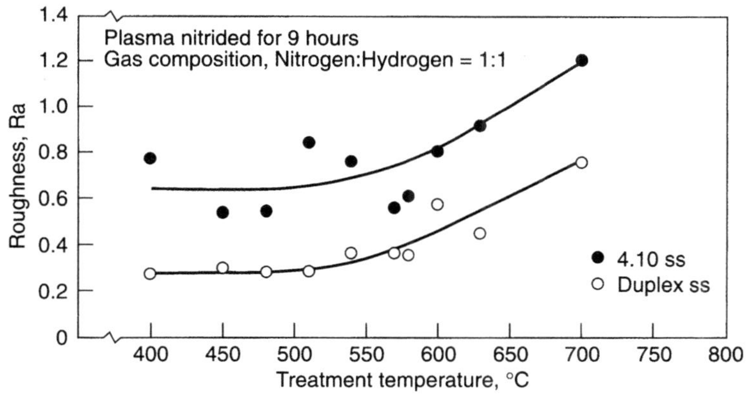

Fig. 8. Roughness values of plasma nitrided steels as a function of treatment temperatures.

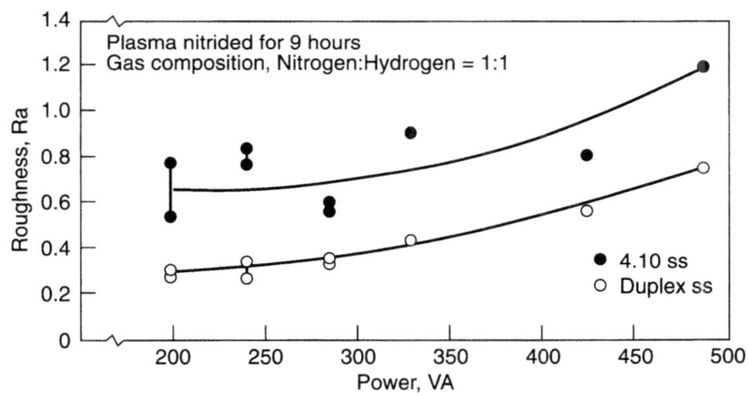

Fig. 9. Roughness values of plasma nitrided steels as a function of the power input.

The roughness effect of treatment time up to 16 h at 570°C was observed to be marginal. Treatment temperature was certainly more marked in its effect. This is reasonable as higher temperatures necessarily mean a higher voltage drop and power input. Figures 8 and 9 show the R_a values plotted against treatment temperature and power. More work is presently in progress.

Industrially, roughness specification and inspection are usually carried out more rigorously at the R & D and test-run phases. Thereafter, visual inspection and touch are then employed for superficial inspection.

2.4 Degreasing and Ultrasonic Cleaning

With plasma heat treatment, the workpiece constitute the cathodic electrode of the whole system. Glow discharges are rather intolerant of contaminants. Any grease or foreign particles would most certainly attract all the power energy to concentrate upon them, resulting in excessive sputtering which would lead to extensive arcing. Hence, thorough degreasing, and, where necessary, ultrasonic cleaning would be most beneficial to ensure uniform treatment. Under normal circumstances, vapour degreasing should be sufficient.

2.5 Scale and Oxide Removal

Scale and oxide removal and derusting are also necessary to ensure quality treatment and an optimal use of power. As plasma nitriding involves sputtering of the surfaces, any oxides or scales are regarded effectively as 'foreign' to the base metal. These would affect the overall thermal and electrical conductivity of the workpiece, rendering a less-than-ideal treatment. Hence, all oxides and scales must be thoroughly removed. This can be accomplished by chemical, mechanical or a combination of both methods.

2.6 Masking and Jigging

Where selective nitriding is desired and considered crucial, masking can be used to protect the surface from the nitriding medium. This can be accomplished by employing masking paint or a physical mask. Masking would invariably involve additional secondary preparation. This will incur costs and therefore to be avoided wherever possible. Figure 10 shows a pair of plasma nitrided automotive shafts with its bottom half masked to avoid the nitriding medium.

Good jigging is also essential where high volume production is concerned. For

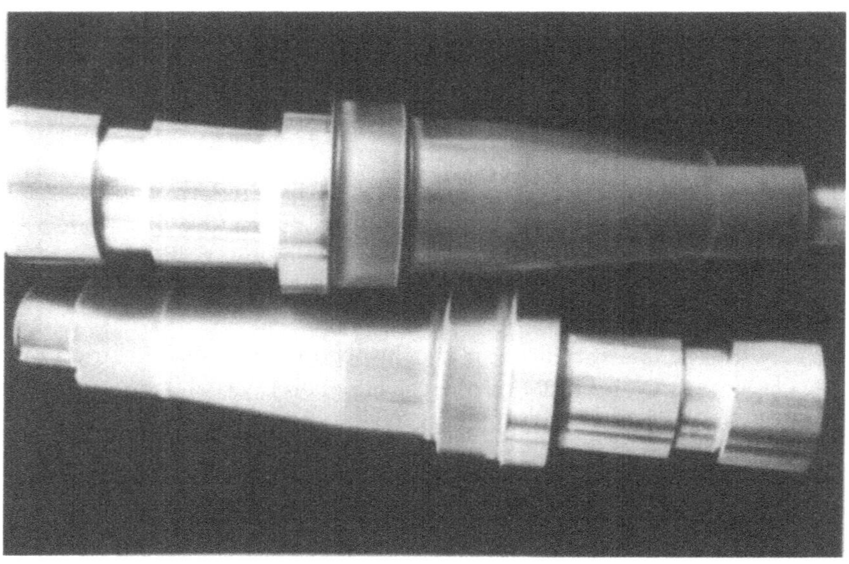

Fig. 10. Plasma nitrided automotive shafts, with bottom halves masked from the nitriding medium.

economical reasons, the maximum number of parts that can be treated in one load must be faithfully calculated along with the jigging to ensure uniform and consistent treatment.

3. PLASMA NITRIDING PROPER

Whilst it is true that case uniformity is ensured by the uniform glow seam which envelopes the workpieces completely, it must be emphasised that temperature differences can arise as a result of changes in the surface area versus volume ratio. As most engineering components are unlikely to be of simple geometry, it goes without saying that the operators must now decide the most important reference point from which to monitor the whole process. Ideally, specimens of similar sizes and chemical composition should only be loaded at any one time to ensure thermal uniformity. Where this is impossible, or uneconomical, then compensation must be allowed for the differences in specific mass.

Unlike the conventional heat treatment processes where the heating elements and the workpieces are separate entities, the workpiece(s) in a plasma nitriding furnace constitute its cathodic electrode. Hence, it is crucial that the load per run must be intelligently calculated to ensure quality treatment. In this instance, both the mass and surface areas must be catered for to ensure sufficient energy input for the required temperature.

Where blind holes and key ways are concerned, it is of paramount importance that the hollow cathode effects are not excessive or this would lead to gross distortion and uneven treatment.

3.1 Temperature Measurement

Industrially, temperatures are measured either by conventional thermocouple or/and infra-red pyroscope. The Germans and French tend to prefer the thermocouple for monitoring temperature and use the pyroscope for inspection on temperature uniformity. The Japanese, on the other hand, carried out a considerable amount of R & D on infra-red pyrometry and use it for temperature monitoring.

With the sheathed thermocouple, it is crucial that the thermocouple is electrically isolated from the cathodic workpiece or the temperature recorded will not be accurate. Hence, regular visual inspection is necessary, to ensure no glowing of the thermocouple. The site and depth of insertion of the thermocouple is also critical to ensure effective thermal representation of the workpiece. Figure 11 shows an example of termperature deviation as a function of gas pressure and depth of insertion.[3]

With the infra-red pyroscope, its main advantage must surely lie in the fact that it measures the surface temperature per se. Much research has gone into this technology; and the Japanese have acquired considerable know-how to achieve temperature accuracy up to $\pm 5°C$.

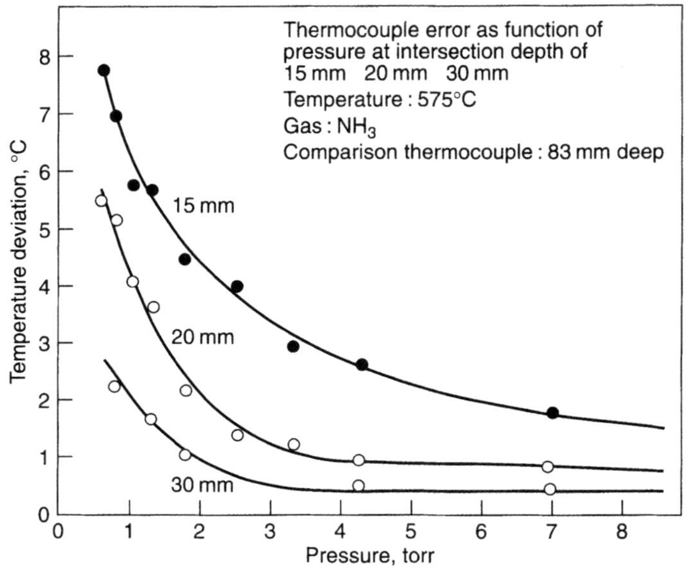

Fig. 11. Temperature deviation as a function of pressure and insertion depth of thermocouple.[2]

3.2 Automation

Programmed automation control on processing parameters such as gas pressure as a function of temperature and power input renders the plasma nitriding process rather 'user-friendly'. This, however, is not without a price, as operators are then inclined to leave all judgement to the 'smart machine'. This can be disastrous in some instances, bringing about considerable damage to the workpieces.

4. POST-TREATMENT INSPECTION

Acceptance procedures are usually specified by the respective clients. Where these are not forthcoming, then it is recommended that any combination or all of the six procedures described below be used for quality assurances.

4.1 Visual Inspection

Visual inspection is perhaps the quickest but most easily ignored procedure. Yet, if performed meticulously and conscientiously, it can reveal much information for the operator. A low magnification glass of up to 20× is usually most helpful; in particular, when inspecting to discern microcracking, heat check and dent.

In a visual inspection on routine basis, the primary interest would be to detect any arc damage and irregularity of surface finishes. The former would suggest poor or incomplete preparation technique whilst the latter would indicate gross variation in temperature or insufficient gas flow.

4.2 Superficial Surface Hardness Measurement

All too often, technicians and engineers tend to forget that the nitriding case depth is less than 0.5 mm; and there have been cases where standard HRC measurements were made for quality control! It must be remembered that any direct surface indentation on a plasma nitrided workpiece will measure effectively the composite hardnesses of the compound layer, the diffusion layer and the matrix. Hence, the reading so obtained is at best indicative and not to be taken absolutely. This is true even if the superficial surface hardness measurement is carried out on HR15-N and HR30-N scales.[4]

To all intents and purposes, then, the scratch or file test would be a good alternative in providing some superficial check on the quality of treatment.

4.3 Core Hardness Measurement

Where the retention of core hardnesses is important, then it must be verified that no softening has occurred. To this end, the untreated surface should be used for evaluation. Depending on the size of the workpiece, the appropriate hardness tester is then chosen for such evaluation – HRC or Vickers would be the likely choice.

4.4 Microhardness Profile

Evaluation of the case depth can best be performed using either the Vickers or Knoop Microhardness tester on metallographically prepared but un-etched specimens. This is perhaps the most useful test for the inspection of case depth in order to ascertain compliances to nitride depth specification.

Some clients may want a complete profile to be plotted out, while others only require microhardness measurements at specified depths below the surface. The latter does help to reduce cost and is usually practised for established treatments. Where a complete profile is required, this entails a thorough survey of microhardness measurements to be taken from close to the surface right down to the matrix.

It goes without saying that the microhardness profile inspection is most crucial during the trial run phase. Definition of case depth, either by a fixed microhardness value or in reference to the core hardness, must then be decided and agreed upon. The load to be applied must be such as to ensure well-defined indentation without radial or circumferential cracking; and all details such as calibration check, inter-indentation spacing and specimen thicknesses must be in compliance to the respective relevant standards.[5,6]

4.5 Metallographic Inspection

Where it is called for and particularly during a test load, metallographic inspection will be most useful. This involves the sectioning of various parts of the plasma nitrided workpiece to study if there is any variation in the thicknesses of the compound layer and case depth.

Good sample preparation is crucial, and all standard and sound metallographic practice must be observed. Etching with suitable agent such as 4% nital or 3% picral plus 1% benzalkonium will darken the case but not the iron nitride compound layer. This will appear white, rendering its direct measurement by microscope rather easy.

With the case depth, however, metallographic inspection should be used only for qualitative evaluation particularly where low alloy and carbon steels are concerned. This is because the case-core interface would not be sufficiently distinctive for optical definition. Whilst the high alloy steels and stainless steels may provide a reasonably well-defined case–core interface, it is nevertheless prudent to use it only qualitatively.

4.6 Dimensional Inspection

Dimensional check of treated parts have been carried out by various methods ranging from the simple micrometer screw gauge, profile projector for linear and angular measurements, roundness testing machine for concentricity to the sophisticated Coordinate Measuring Machine. Insofar as Doxon is concerned, dimensional check has always been evaluated by the clients.

ACKNOWLEDGEMENT

The authors would like to express their gratitude to Doxon Engineering Pte Ltd, Singapore, for all the industrial-related information that was necessary for the writing of this paper. Appreciation also goes to the Nanyang Technological University for the research work and data, without which this paper could not have been prepared.

REFERENCES

1. *Singapore Invedstment News*, Precision Engineering Supplement, November 1993, 14.
2. M. Bader, F. C. Witteborn and T. W. Snouse: *NASA Report: TR-R-105*, 1961.
3. Operational Manual from Klockner Ionon GmbH.
4. ASTM E18-92: Rockwell Hardness and Rockwell Superficial Hardness of Metallic Materials.
5. ASTM E384-84: Microhardness of Materials.
6. JIS Z 2251-80: Method of Microhardness Test for Vickers and Knoop Hardness.

SCRATCH ADHESION TESTING AND ITS SIGNIFICANCE FOR COATING PERFORMANCE

S. J. BULL

Materials Division, Department of Mechanical, Materials and Manufacturing Engineering, University of Newcastle, Newcastle-upon-Tyne, NE1 7RU, UK

ABSTRACT

Perhaps the most important property of any coating is its adhesion to the substrate since many functional requirements such as tribological performance are critically dependent on this important property. There is no universal adhesion test which can be applied to all coating/substrate systems but the scratch test can give information about performance in a wide range of systems where other tests (such as the direct pull-off test) are not applicable. As a ranking test for thin coating adhesion the scratch test is particularly useful but the critical load data generated is dependent on a range of extrinsic (i.e. equipment and test-specific) factors and intrinsic factors (i.e. film and substrate properties). It is important to account for both of these effects if the results of the tests are to be applied to understanding coating performance. This paper reviews the factors affecting the scratch test and how these influence the magnitude of the critical loads determined. The relationship between scratch test results and performance for thin and thick coatings is discussed.

1. INTRODUCTION

No engineering adhesion test is able to measure the basic adhesion of a coating to its substrate, i.e. the interfacial bond strength. Experimental 'adhesion strengths' (EA) are dependent on basic adhesion (BA) but also on other factors specific to the test method such that

$$EA = BA \pm EXT \pm INT \quad (1)$$

where EXT are extrinsic parameters such as the equipment configuration, test parameters and ambient atmosphere, and INT are intrinsic properties of the coating/substrate system such as the mechanical properties of coating and substrate, coating thickness and roughness. Thus, in order to fully determine the relationship between basic adhesion and any experimental test results, it is important to carefully control both the EXT and INT factors. EXT factors can be controlled by rigidly maintaining a well defined test protocol whereas the effect of the INT factors requires very careful calibration to determine the functional dependency of experimentally measured adhesion on critical properties of the coating/substrate system. This can be difficult to achieve and makes comparisons between very different coating/substrate systems problematic.

Several reviews have surveyed the range of adhesion tests available for coatings[1–6] and from these the scratch test can be considered to be the most widely applicable because it is relatively simple to perform, has been successfully used to determine the effects of deposition parameters and can give information in cases where no other test can be used. The test essentially exists in two modifications:

(1) Thin films (generally $<20\,\mu m$) – in this case a scratch diamond (usually a Rockwell C diamond with a 200 μm radius tip) is drawn across the coated surface under ever increasing loads until some well defined failure occurs at a load which is called the critical load, L_c. This failure may or may not be related to adhesion depending on the system under investigation but can be used to rank the scratch resistance of the coating which is important for tribological performance in some applications. If the failure does depend on adhesion then the critical load can be used as a semi-quantitative measure of experimental adhesion.
(2) Thick films (generally $>30\,\mu m$) – in this test a polished cross-section is produced by standard metallographic techniques and a fixed-load scratch is performed starting in the substrate and passing through the coating into the mount (or free space). The process is repeated until, at some critical load, the stylus drags the coating from the substrate and opens up an interfacial crack. Plotting the length of this interfacial crack against load can be used to fix the critical load more exactly. This test is less well developed than the thin film scratch test but is showing some promise for the assessment of plasma sprayed coatings.

In this paper the use of each modification of the scratch test to determine coating adhesion (or some other performance criterion) will be discussed and the factors which influence the test introduced. The significance of scratch test results for the assessment of coating performance will then be discussed in light of these factors.

2. THIN FILM SCRATCH TESTS

In the thin film scratch test a diamond stylus drawn across the coating surface induces stresses on the coating, substrate and the coating/substrate interface which can lead to a range of failure modes that may be relevant to coating performance. If the failure mode depends on adhesion then the scratch test can be used to give a semi-quantitative measure of adhesion provided that the onset of a particular failure mode can be unambiguously established for each coating/substrate system under investigation.

Given a new system to investigate a number of stages are necessary before significant results can be produced:

(1) Establish the failure mode.
(2) Fix the test protocol (to control extrinsic variables).
(3) Establish the effects of intrinsic variables or control over substrate and coating properties.
(4) Determine the point to point variability.

Once this procedure has been completed it is possible to define an unambiguous critical load criterion, corrected for intrinsic effects, which can be used to assess coating performance. The following sections discuss some of these stages in more detail.

2.1 Failure Modes

A number of failures can occur during the scratch test ranging from plastic deformation or fracture of substrate (or coating) to coating detachment.[7–8] There may be several mechanisms by which each failure mode can occur (e.g. spallation can occur both in front of or behind a moving scratch diamond) so it is important to unambiguously determine a well-defined failure mode with an understandable cause which can be easily recognised. Figure 1 shows a schematic of the failure modes

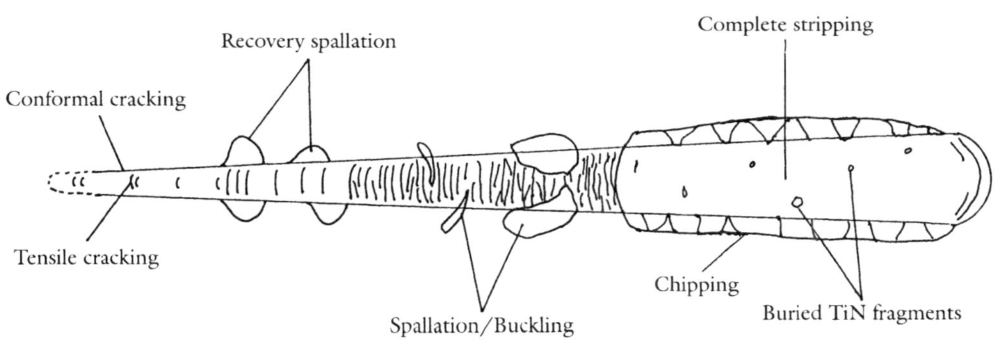

Fig. 1. Schematic of the failure modes associated with an increasing load scratch for a 2 μm TiN coating on M2 steel.

association with a scratch track for a 2 μm TiN coated M2 steel coupon. At low loads a visible scratch track is produced by plastic deformation of the substrate and, as the coating bends into the track, some cracking occurs at the edge of the tracks. Two types of spallation can be observed at the edge of the track at higher loads. Recovery spallation occurs in response to differential elastic recovery after the scratch diamond has passed and starts at the cracks at the track edge. This leads to spalled regions beside the track with no apparent damage to the track itself. Spallation or buckling due to the compressive stresses generated ahead of the moving indenter can also occur and usually associated with damage within the track as well as outside it. This generally occurs at higher loads than recovery spallation and is observed for many more coating/substrate systems. When the coating is very well adhered or the coating surface is very rough it can be difficult to resolve this type of failure, particularly when through-thickness cracking within the track (either tensile or conformal) complicates matters. In this case there is a need for very careful microscope analysis of the scratch track after scratch testing.

The most reliable method to determine failure will always be post facto microscopic analysis by reflected light microscopy or scanning electron microscopy (SEM). Only by this method can adhesion-related failures be unambiguously detected. However, critical loads can also be determined in terms of other signals which can be measured during the scratch test.

The simplest to interpret is the friction force measurement[9] which tends to change when the coating is removed or penetrated (Fig. 2). For hard coatings such as TiN on a steel substrate both the magnitude of the friction force and its variability along

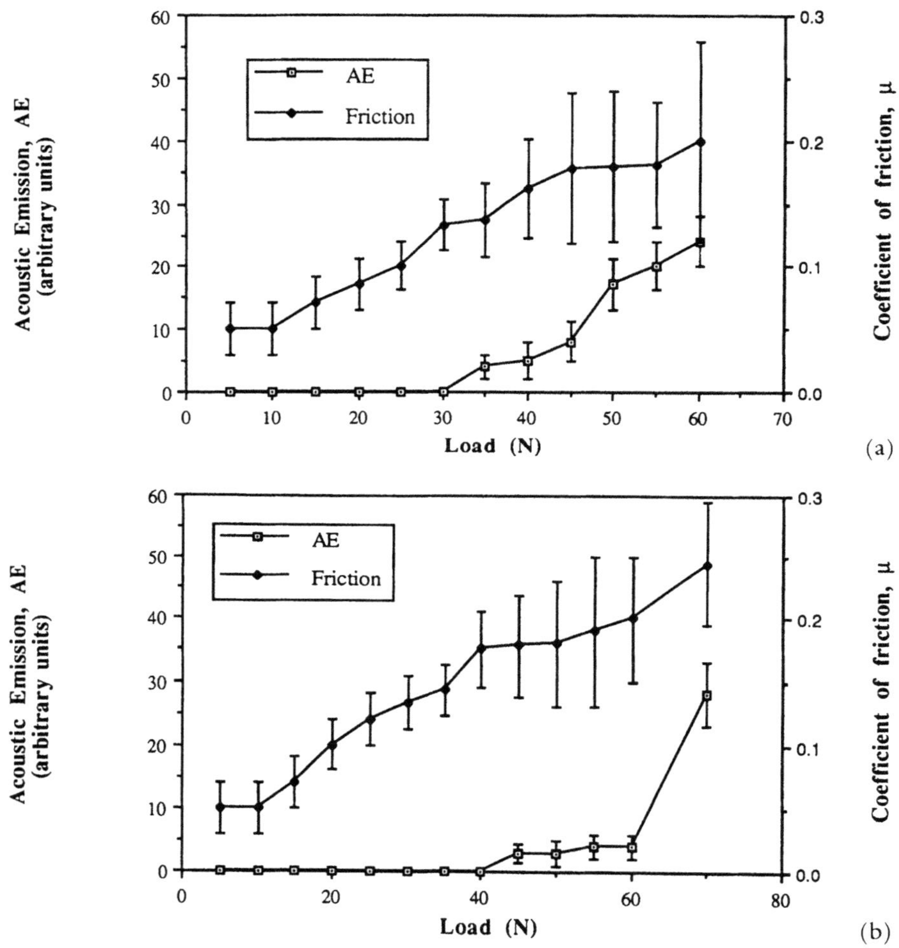

Fig. 2. Friction and acoustic emission signals as a function of load for (a) 1 μm TiN on M2 and (b) 6 μm TiN on M2.

the track will increase. This is a particularly useful method for assessing the scratch adhesion behaviour of very thin coatings ($\leqslant 1\ \mu m$) where resolving individual failures is made difficult by large amounts of through-thickness cracking.

In addition, the measurement of the Acoustic Emission (AE) generated during scratch has been used to determine a critical load.[10] The magnitude of the acoustic emission signal is sensitive to the number and size of cracking events which occur during scratching and contributions from the substrate, through-thickness cracking in the coating, chipping and coating spallation are contained in the signal. However, for some coating/substrate systems a large increase in the measured AE signal occurs at the critical load – if this can be correlated with a spallation event (rather than chipping within the coating or substrate) it can be used to determine a scratch adhesion critical load. The gradual build-up of the AE signal as a function of scratch test load can also be used to determine the relative damage resistance of the coating.

Which failure modes will occur will depend on the properties of the coating and substrate. Soft metal films tend to be plastically deformed during the test and show no cracking or chipping. However, spallation can be observed if the adhesion is very poor. In cases where the adhesion is good the change in friction on uncovering the substrate can be used to determine a critical load. Otherwise careful microscopy or X-ray microanalysis in the SEM is needed to determine if the coating is removed in the scratch track. Clearly, the mechanical properties of the coating and substrate will thus be important in determining the critical load. For hard coatings such as TiN the magnitude of the critical load and the type of failure which is observed depends on the brittle or ductile nature of the substrate.[8] Brittle failure leads to much lower critical loads.

2.2 Extrinsic Test Parameters

There are two basic types of scratch testers which are commonly in use:

(1) Manual dead-load machines. In this case a scratch is made over a fixed length with a fixed applied load. After this a new weight is added, the sample is moved and the process is repeated. It is necessary to perform a number of scratches before L_c can be determined.

(2) Automatic machines. In this case a single scratch is produced using a spring-loaded diamond. The load is continuously increased along the length of the scratch until failure occurs. The test is quicker and easier to perform than the simple manual test but is more sensitive to the larger flaws at the coating/substrate interface.

There is no reason to believe that the absolute critical load values determined by the two tests should be the same without calibration of instrument effects and indeed the critical loads produced by the automatic test have been shown to be somewhat lower than in manual tests on the same samples[11] (Fig. 3) even if the performance trends which are predicted are the same. Even for a single type of tester slight

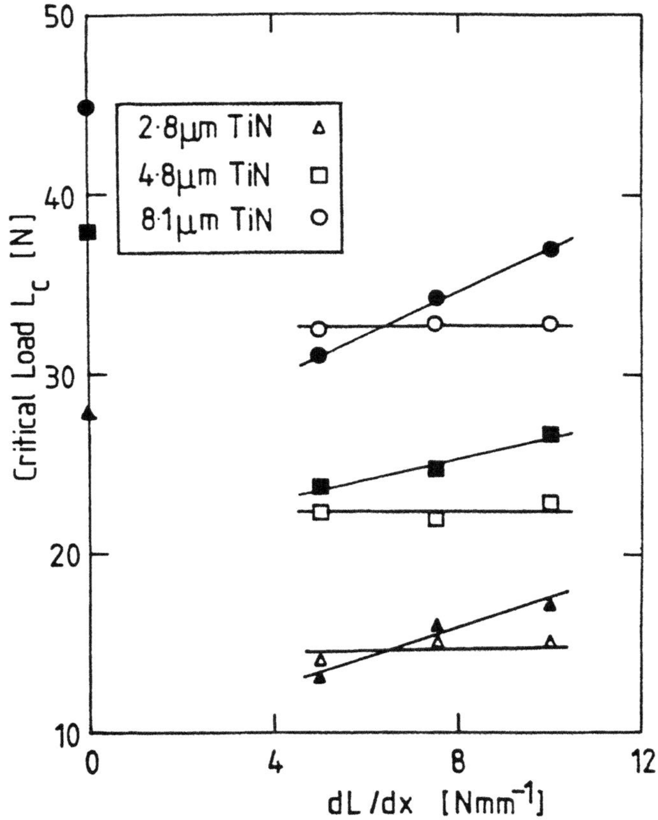

Fig. 3. Variation of critical load with loading rate for TiN coatings on BO1 steel.

variations in the construction can lead to variations in machine stiffness and hence the load at the scratch diamond which can affect the critical load. Clearly, to compare results between different laboratories (or even different machines in the same laboratory) it will be important to calibrate machines using standard coating/substrate systems of controlled adhesion. Such standardisation work is only just beginning and it will be a long time before reliable interlaboratory comparisons will be achievable.

It is therefore necessary to make any performance assessments using a single machine and to devise a test procedure which is fully reproducible. For the manual test the main variables which should be controlled are:

(1) Scratch length (typically 3 mm).
(2) Traverse speed (typically 10 mm/minute).
(3) Diamond stylus (composition, tip radius, orientation, wear).
(4) Contamination (of both diamond and coated surface).
(5) Test environment (composition, humidity).

The machine variables (1) and (2) are easily set and verified and a solvent cleaning procedure similar to that used for preparing substrates for coating is adequate to remove most contamination for (4). Similarly a well appointed test laboratory will have a good controlled test environment and daily variations generally make only a small difference to the measured critical load values. It is therefore the state of the scratch diamond which is critically important and needs to be carefully monitored if reliable comparisons are to be made. As a scratch diamond wears the critical load value determined using it is progressively reduced. Similarly the diamond can pick up debris from the substrate surface which is not easily removed by solvent cleaning or wiping. If the diamond is used to test another sample without proper cleaning (e.g. by etching or polishing) an anomalously low critical load will be recorded. Prior to testing any new substrate the diamond should be carefully inspected for signs of wear (cracking, scoring, chipping) or debris and rejected if any damage or tenacious debris is visible. Diamonds can be repolished and re-used but it is important that the 200 μm tip radius is maintained to a high degree of accuracy. Ideally, detailed comparisons should be made using the same scratch diamond mounted in the tester in the same orientation, starting at low loads and increasing until the critical load is reached and just passed, since running to very high loads leads to copious debris transfer to the diamond. Any debris is then removed prior to subsequent testing.

In the automatic test the same factors are important but in addition the loading rate must also be controlled. As the loading rate increases there is a tendency for the critical load to increase.[11] This is probably due to damping and compliance effects in the tester arm – although strain-rate sensitivity in the coated material cannot be ruled out.

A typical test procedure would thus be:

- inspect scratch diamond;
- clean diamond and coated coupon;
- fix machine variables (loading rate, etc.);
- perform scratch(es);
- record friction, acoustic emission signals as a function of load;
- record position and direction of scratches to enable comparison with friction and AE signals;
- identify critical load criteria by microscopy, AE or friction;
- normalise results for intrinsic factors if required (see section 2.4).

2.3 Definition of the Critical Load for Adhesion Testing

In the scratch adhesion test using a manual tester the critical load is usually taken as the load at which the coating is removed in a regular way along the whole channel length[2] although, due to the fact that there is a distribution of flaws at the coating/substrate interface, isolated areas of coating removal do occur at lower loads and the critical load transition is somewhat subjective.[12] By counting the number of failures at a given load along a fixed scratch track length it is possible to gain information about the interfacial flaw distribution and define a critical load, L_a, which is tied to the flaw distribution and much less subjectively defined (Fig. 4). In addition, this

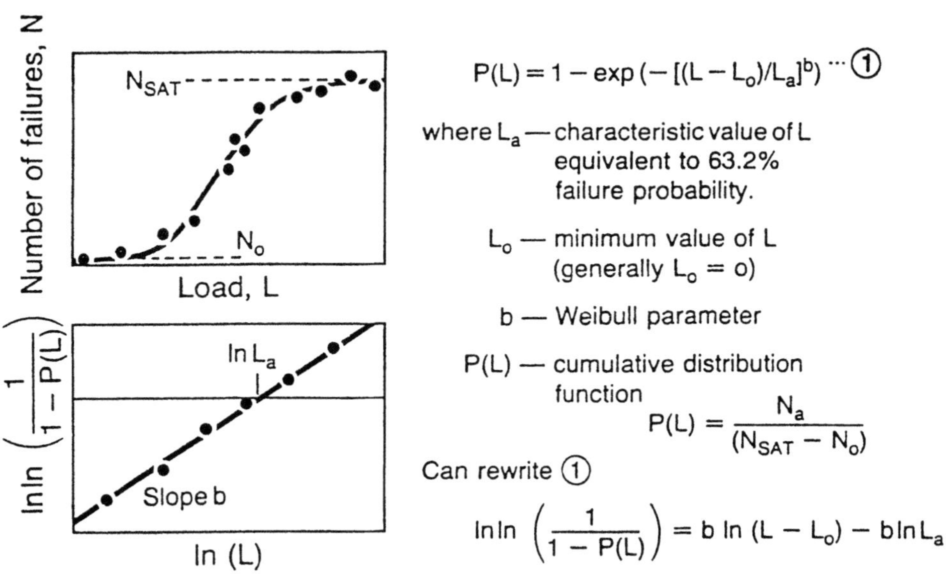

Fig. 4. Weibull statistics analysis of the scratch test.

Weibull statistics approach yields an important parameter, the Weibull Modulus, *m*, which is a measure of the variability of coating adhesion. The cumulative failure probability, $P(L)$ is defined as

$$P(L) = \frac{N(L) - N(0)}{N_{SAT}} \tag{2}$$

where $N(L)$ is the number of failures at load L, $N(0)$ is the number of failures at low loads (usually $N(0) = 0$) and N_{SAT} is the saturation number of failures, or the maximum number of resolvable failures in cases where this is not reached. $P(L)$ can be given by the Weibull distribution

$$P(L) = 1 - \exp\left[-\left(\frac{L - L_o}{L_a}\right)^m\right] \tag{3}$$

where L_o is the load at which the probability of failure is zero (generally $L_o = 0$) and L_a is a constant equal to the load at which there is a 63.2% cumulative failure probability.

Using this approach, it has been shown that TiN coatings produced by sputtering and plasma-assisted CVD (PACVD) have very similar critical loads but different interfacial flaw distributions. The broader interfacial flaw distribution in PACVD TiN (produced by chlorine contamination) leads to the relatively poor tribological performance of this coating.[13] The Weibull statistics approach can also be applied to other failure modes which do not depend on adhesion. It can also be applied to the acoustic emission signal if the average AE signal along the track is used to replace the number of failures in equation (2).

For the automatic scratch test where the load increases continuously along the scratch track the load at which failures are regularly observed along the scratch can be determined by measuring the length of track before this occurs. Isolated low load failures can still occur, particularly at scratches or machining marks on the sample surface. For this reason it is important to perform at least three tests and carefully examine the tracks to determine the critical load reliably. No information about interfacial flaw distributions can be determined in this test, but it is relatively easy to relate large pulses of acoustic emission with individual failures along the track due to the increased size of failures produced as the increasing load puts extra energy into the system compared to the manual dead-loaded case.

2.4 Intrinsic Factors

There are a range of coating/substrate systems parameters which are known to affect the critical load:

- coating thickness;
- substrate hardness;
- coating hardness;
- residual stress in coating;
- diamond/coating friction;
- surface roughness.

The hardness of the coating and substrate in turn depend on the yield strength and Young's moduli of the materials. In laboratory testing many of these factors can be fixed so as to determine the variation of the critical load with a single parameter. However, in real applications several factors can vary at the same time and this makes meaningful comparisons difficult. At present there is no good theoretical model of the scratch test which can take these factors into consideration and, given the complexity of the stress state beneath a moving indenter on a coated system, it is difficult to see how one could easily be developed. Thus a semi-empirical approach is needed to enable scratch tests to be normalised when several intrinsic factors vary from run to run (as in most commercial coatings).

The following sections describe the critical load variations observed with the main intrinsic parameters.

2.4.1 Substrate Hardness

For both brittle and ductile materials the critical load increases with substrate hardness[3,4,14] as the substrate gives increasing load support to the coating (Fig. 5). In most cases the increase is linear; for sputter ion plated (SIP) TiN on a range of substrates

$$L_o = L_o + k_1 H_s \tag{4}$$

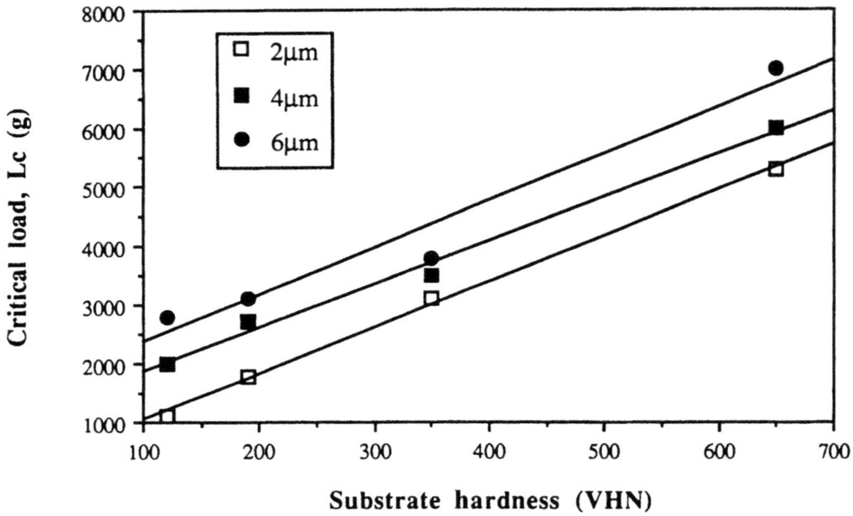

Fig. 5. Variation of critical load with substrate hardness and thickness for sputter ion plated TiN.

where $L_o = 2.1$ N and $k_1 = 8.3$ N/GPa. A typical tool steel substrate might range in hardness from 8.5 to 11 GPa, giving a variation in critical load of 71 to 94 N.

2.4.2 Coating Hardness

Very soft coatings are often plastically deformed in the scratch test and are thinned to transparency rather than removed by spallation in the test. Indeed, coatings which can accommodate contact stresses by plastic deformation do not show any signs of removal during the test. As the hardness of the coating increases there is a transition away from plasticity dominated behaviour to elastic/plastic behaviour which is characterised by the observation of fracture as well as plasticity along the scratch track. Spallation is now observed. As the coating becomes harder and increasingly more brittle the critical load is reduced. Again, an approximately linear behaviour is observed for SIP TiN on stainless steel.

$$L_c = L_o - k_2 H_c \quad (5)$$

where $L_o = 38$ N and $k_2 = 0.6$ N/GPa. Typical TiN coatings vary in hardness from 23–28 GPa, giving a variation in L_c from 21 to 24 N. The variation of L_c with coating hardness is thus relatively small. In practice this behaviour is also influenced by the residual stress in the coating.

2.4.3 Coating Thickness

The critical load has been observed to increase with coating thickness by a number of workers.[4,14,15] The increase often appears to be linear for some substrates but can appear to saturate at the highest coating thicknesses for materials such as stainless or

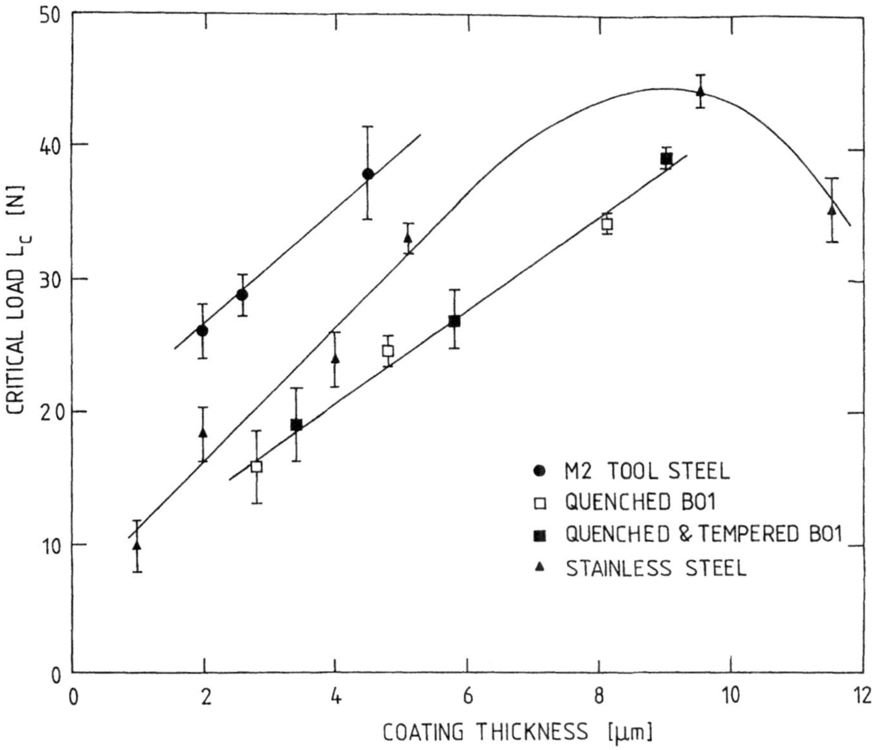

Fig. 6. Variation of critical load with coating thickness for sputter ion plated TiN on several steel substrates.

tool steels (Fig. 6). In this case the critical load is approximately proportional to the square root of coating thickness.

$$L_c = L_o + k_3 \sqrt{t} \tag{6}$$

where for SIP TiN on stainless steel $L_o = 12.3\,\mathrm{N}$ and $k_3 = 1.26 \times 10^4\,\mathrm{Nm}^{-1/2}$. TiN coating thicknesses typically vary from 1–10 μm with a predicted variation in critical load from 25–52 N. Clearly, coating thickness has a large effect on the measured L_c value. At very low coating thicknesses (<1 μm) for TiN films there is a slight decrease in critical load as coating thickness increases. This is due to changes in the observed failure modes since very thin coatings do not generally produce appreciable spallation whereas this is much more apparent for thicker layers (>2 μm).

2.4.4 Coating Roughness

As the surface roughness of the coating increases the critical load is reduced.[4,11] Since most thin coatings replicate the roughness of the substrate this can be directly influenced by the roughness of the substrate prior to coating. An approximately log linear variation is produced for 2 μm SIP TiN coatings on M2

$$L_c = L_o - k_4 \log R_a \tag{7}$$

where $L_o = 31$ N and $k_4 = 20$ N. Typical engineering components vary in roughness from 0.1–5 μm R_a giving a variation in L_c from 45–51 N. This is a relatively minor variation in critical load. For many systems if R_a is less than 100 nm there is no variation of critical load with roughness. However, such highly polished surfaces are rarely observed on engineering components.

2.4.5 *Residual Stress*

As the residual stress in the coating increases the critical load is reduced for most coatings since the elastic strain energy stored in the coating due to the residual stress will add to that induced during scratching.[16] For SIP TiN the variation is approximately linear

$$L_c = L_o - k_5\sigma_c \tag{8}$$

For SIP TiN on stainless steel $L_o = 33$ N and $k_5 = 1.7$ N/GPa. Residual stresses tend to vary from 2–10 GPa, giving a variation in L_c from 16–30 N. This can be a large effect. It is difficult to separate this effect from that due to the hardness of the coating (equation (5)).

2.4.6 *Diamond/Coating Friction*

The shear stresses on the coating/substrate interface which lead to coating detachment depend on the magnitude of the surface traction on the coating/substrate system induced by the scratch diamond. This is controlled by the friction between the diamond and the coating surface. If this is very low the diamond slides across the surface without coating detachment as the maximum shear stress generally occurs well below the surface leading to plastic deformation of the substrate. However as μ increases the maximum shear stress moves closer to the surface until it can act on the coating/substrate interface (typically when $\mu \geqslant 0.25$). Detachment readily occurs when these conditions are met. The critical load for SIP titanium nitride on stainless steel coated with a range of metals to modify the diamond/coating friction was found to be inversely proportional to the friction coefficient in cases where the metal film was not removed prior to TiN failure.[17] For SIP TiN on stainless steel (2 μm thick)

$$L_c = \frac{k_6}{\mu} + L_o \tag{9}$$

where $L_o = -4.9$ N and $k_6 = 4.7$ N. In the scratch test the coefficient of friction typically varies from 0.1–0.3, giving a variation in L_c from 11–42 N. Clearly, friction is very important in dictating the critical load.

2.5 Normalisation of Intrinsic Parameters

The intrinsic parameters discussed in the previous sections fall into two distinct groups:

(1) Those which have a large effect on L_c (coating thickness, substrate hardness, and diamond/coating friction).

(2) Those where the effect is small except in exceptional cases (coating hardness, residual stress, surface roughness).

If the effects of intrinsic variables can be considered independent, equations (4)–(9) can thus be rewritten

$$L_c = L_o + k_1(H_s - H_s^0) - k_2(H_c - H_c^0) + k_3(\sqrt{t} - \sqrt{t_0}) - k_4 \log\left(\frac{R_a}{R_a^0}\right) - k_5(\sigma_c - \sigma_c^0) + k_6\left(\frac{1}{\mu} - \frac{1}{\mu_0}\right) \quad (10)$$

where the subscript 0 denotes a fixed value for comparison. Suggested values are $H_s^0 = 10$ GPa, $H_c^0 = 25$ GPa, $t_0 = 1$ µm, $\sigma_c^0 = 4$ GPa, $R_a^0 = 0.05$ µm and $\mu_0 = 0.2$. L_o is then the normalised critical load corrected for intrinsic factors.

Table 1. Measured and normalised scratch test critical loads for a range of PVD TiN coatings.

Process	Substrate	Thickness (µm)	Substrate hardness (GPa)	L_c measured (N)	L_o normalised (N)
SIP	Stainless steel	1.5	1.8	10	88
		2	1.8	18	75
		2.9	1.8	20	72
		4.1	1.8	23	71
		5.5	1.8	32	85
		9.5	1.8	42	92
		11.5	1.8	35	82
SIP	M2	2	6	26	32
		3.8	6	32	36
		4.5	6	37	43
		6.5	6	52	54
		9	6	60	61
Arc evaporation	Stainless steel	1.7	1.8	9	69
		3.2	1.8	14	73
		4.8	1.8	19	77
Magnetron sputtering	Stainless steel	5	1.8	34	79
		5	1.8	34	87
		5	1.8	30	86
		5	1.8	30	71
		5	1.8	32	71
Ion beam assisted deposition	Stainless steel	1.4	1.8	15	73
Ion plating (PAPVD)	Stainless steel	0.9	1.8	10	84
Ion plating (PAPVD)	M2	1.36	8.8	38	44
		1.98	9.5	30	39

Table 1 shows the measured and normalised L_c values for a range of Physical Vapour Deposition (PVD) TiN coatings on different components. Although there is some scatter in the critical loads determined even within a given process the normalised values are much more consistent. The normalised L_o values imply that the adhesion for most of the coatings is very similar on a given substrate. The normalised adhesion of coatings on M2 is lower than that of stainless steel. This is due to a different operating failure mode (recovery spallation rather than buckling/spallation).

In Table 2 the measured and normalised critical load values for Plasma Assisted Physical Vapour Deposited (PAPVD) coatings of a range of commercial hard coating materials on tool steel are presented. TiCN and CrN have better adhesion to M2 compared to TiAlN and TiN which are very similar. In the case of CrN this is due to differences in failure mode but it appears that the adhesion of TiCN is better than the other coatings. However, much more data are needed to establish the validity of this normalisation approach.

Table 2. Measured and normalised scratch test critical loads for a range of PAPVD hard coatings on M2.

Coating	Thickness (μm)	Substrate hardness H_s (GPa)	R_a (μm)	μ	Residual stress (GPa)	Coating hardness H_c (GPa)	Measured L_c (N)	Normalised L_o (N)
TiN	1.36	8.8	0.16	0.119	6.0	26	38	44
	1.98	9.5	0.27	0.146	6.0	26	30	39
	1.31	9.2	0.24	0.117	6.0	26	25	31
	1.31	9	0.24	0.1	6.0	26	17	18
	2.04	8.8	0.06	0.16	6.0	26	42	46
TiCN	2.14	10	0.3	0.172	6.0	40	59	77
	3.15	9	0.13	0.17	6.0	40	45	60
TiAlN	2.6	8.7	0.11	0.16	6.0	26	41	49
CrN	3.14	7.9	0.07	0.167	6.5	22	55	63

2.6 Scratch Test and Performance

The scratch test can be effective at assessing performance in those applications where the loading geometry and failure modes are similar to the test. Coatings for machine parts undergoing mild abrasive wear are often well ranked by the scratch test.

For tool coatings the behaviour is not so easy to rationalise. The scratch test can be used as a process monitoring tool to detect batch-to-batch variations in coating performance with a great deal of success.[18] However, relating the scratch test performance to the performance of a particular tool is considerably more difficult. Figure 7 shows plots of measured and normalised critical load against number of holes drilled for a range of TiN coatings deposited onto 6 mm twist drills by PAPVD. Although there is an approximate correlation between drill performance and the critical load measured in the scratch test (an increase in critical load

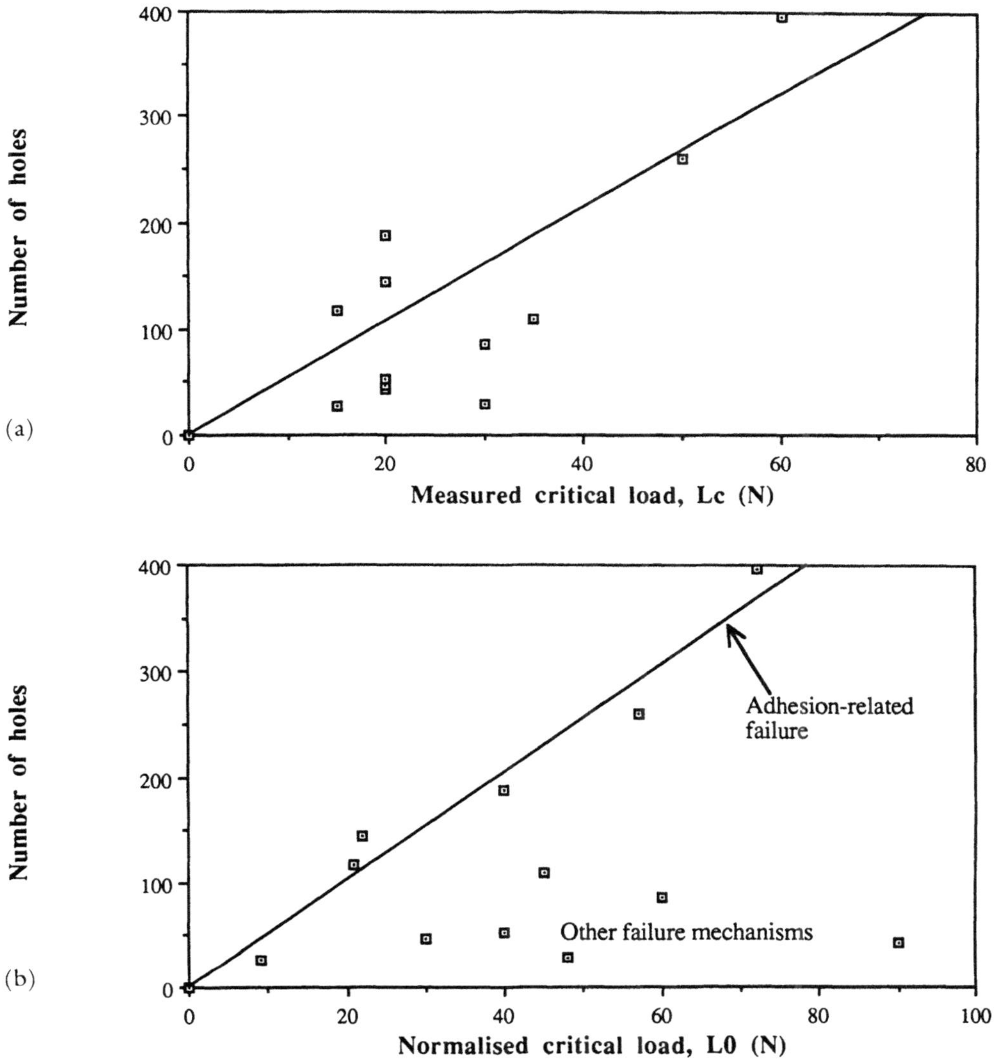

Fig. 7. Variation of number of holes drilled by a 6 mm twist drill in a tyre steel with (a) measured critical load and (b) normalised critical load.

correlates with an increase in drill life), there is considerable scatter in the results. After normalisation results can be divided into two categories, those where the number of holes drilled is proportional to the normalised critical load (which implies that drill failure depends on coating adhesion) and those with poorer drill performance where other failure modes are important.

Analysis of the failed drills indicates that substrate fracture, chemical wear, overheating and transfer of workpiece material to the drill surface can dictate the life of the tools as well as adhesion, wear resistance and coating fracture (which can be assessed by the scratch test). Thus, to provide complete information about the performance of a coated component, the scratch test can only be part of any evaluation programme. However, if achieving a certain minimum level of adhesion is necessary (often the case in tooling applications) the scratch test can be used to give confidence that an acceptable level has been achieved.

3. THICK FILMS

Many adhesion tests, such as the direct pull-off test, cannot be used to assess the adhesion of thick plasma sprayed coatings due to the limitations of the strengths of available adhesives or the requirements for complex sample geometries and test procedures. The scratch test can be used to rank the adhesion of a range of thermally sprayed coatings including some deposited by processes, e.g. High Velocity Oxy Fuel (HVOF), which cannot be easily assessed by other techniques.[19,20]

The test involves traversing a loaded diamond indenter across the polished cross-section of a sample. The damage introduced (Fig. 8) can be used to quantify both the adhesion of the coating and its cohesive strength. As the scratch diamond exits the coating surface a conical plug of material is removed, the size of which is inversely related to the cohesive strength of the coating. Tests on thermally sprayed tungsten coatings show how the techniques can be used to identify subtle differences in cohesive strength resulting from different deposition conditions as well as the

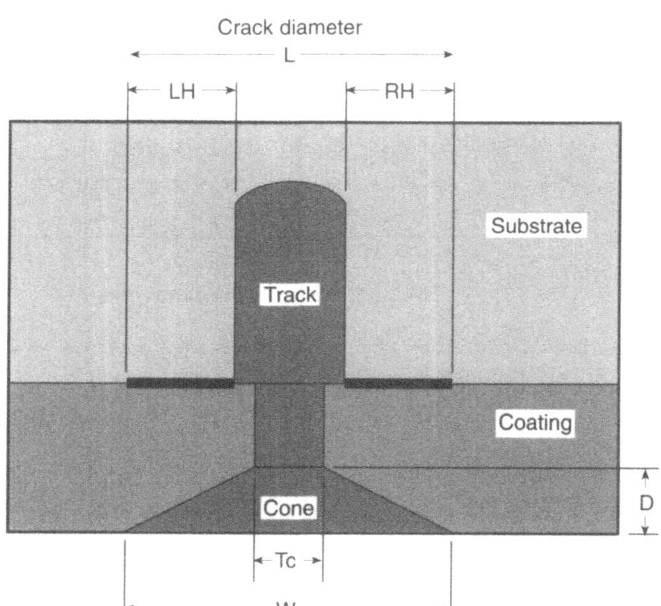

Fig. 8. Schematic of the thick film scratch test.

larger variations due to oxide entrapment (in Air Plasma Spraying (APS) but not Vacuum Plasma Spraying (VPS).[20]

In some cases an interfacial crack of length *I* is produced which can be used to assess adhesion in a similar manner to the cohesive strength. As the scratch diamond approaches the interface from the substrate side it will encounter the edge of the coating and a force will be applied to the coating equal to the friction force. At this point a number of things can happen:

(1) The coating can plastically deform.
(2) The coating can fracture and a plug is removed extending to the interface.
(3) An interfacial crack can be formed to release the elastic strain energy in the coating.

After this the diamond will pass across and out of the coating. For hard coatings mechanism (3) occurs if the coating is thick enough and has sufficient cohesive strength. The interfacial crack does not form until the diamond passes from the substrate into the coating. The tensile stresses behind the moving indentor are thus responsible for failure.

The scratch adhesion test for thick films thus involves making a number of fixed load scratches and measuring the size of the interfacial crack as a function of applied load. Extrapolation lines for each coating back to zero crack length give a measure of the critical load for interfacial failure, L_c. This is analagous to the thin film scratch test value.

Figure 9 shows the variation of interfacial crack length with load for three different WC/Co coatings on titanium sprayed by APS, VPS and HVOF. The critical loads

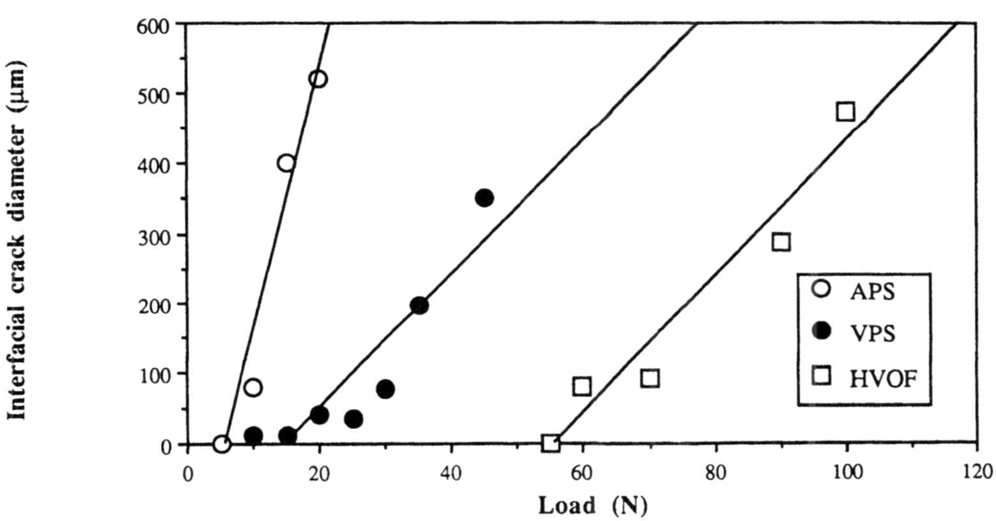

Fig. 9. Variation of interfacial crack diameter with load for WC/Co coatings produced by a range of spraying processes.

Table 3. Parameters obtained from the thick film scratch testing of sprayed WC/Co coatings on titanium.

Deposition technology	Critical load L_c (N)	Slope of crack-length plot (μm/N)
APS	5.9	37.8
VPS	14.9	9.5
HVOF	55.8	9.7

for the onset of cracking increase on going from APS to VPS to HVOF (Table 3). This correlates with the increase in particle speed in the spraying processes. The APS coatings give a steep slope which is indicative of poor adhesion whereas the VPS and HVOF coatings give a shallow slope and the best adhesion. In tribological tests and in a range of applications the HVOF material has outperformed coatings deposited by the other techniques. These results show that this is probably due, at least in part, to superior adhesion.

The magnitude of the interfacial crack depends on the method of mounting of the cross-section. Figure 10 shows a comparison of the crack lengths for the same alumina coating plasma sprayed onto copper mounted in bakelite (hot mount), castable epoxy (cold mount), and unmounted. As the strength of the mount increases the size of the crack is reduced and thus longer cracks are created in the unmounted sample. Clearly the same mounting technique must be employed for all samples which are to be compared.

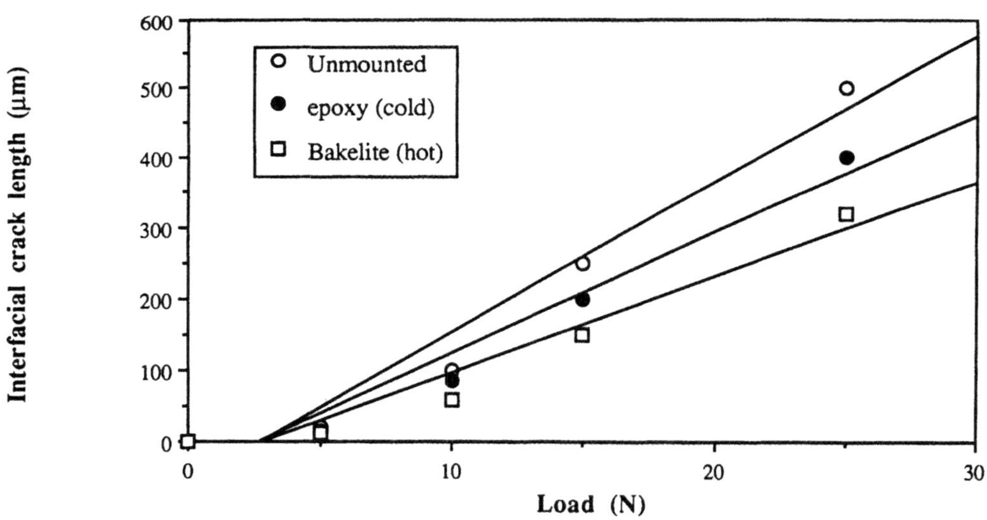

Fig. 10. Effect of mounting medium in the variation of interfacial crack diameter with test load for alumina coatings deposited on copper by air plasma spraying.

3.1 Factors Affecting the Test

As with the thin film scratch test, there are a number of extrinsic and intrinsic factors which affect the test but to date there has been relatively little work done in quantifying these. The following sections discuss some of the factors in more detail.

3.1.1 Extrinsic Factors

The main extrinsic factors specific to the thick film scratch test are:

- traverse speed (typically 10 mm/min);
- loading geometry/method;
- limit of diamond displacement in the vertical direction;
- diamond stylus wear;
- test environment;
- sample preparation.

It is important that the diamond does not cross the coating too quickly as it can jump or cause chipping which will mask any interfacial failure. Controlling the diamond displacement in the vertical direction is important as this can affect the size of cone produced by cohesive failure. This is particularly important if constant load scratches are to be produced using the automatic increasing load scratch tester. The thick film scratch test is less susceptible to the test environment and wear of the scratch diamond but it is critically dependent on the metallographic procedures used to produce the cross-section which must be standardised if the test is to be meaningful.

3.1.2 Intrinsic Factors

The main intrinsic factors in the test are the coating thickness, its residual stress and cohesive strength and the friction between the coating and the scratch diamond. As the coating thickness increases it becomes more difficult to detach it from the substrate and the interfacial crack length is reduced. A compressive stress in the coating will, however, make it easier to detach and increases the crack length. The cohesive strength of the coating is important because this often dictates whether coatings fail at the interface at all. If the cohesive strength is low cracking occurs within the coating and no interfacial failure is produced. Strong dense coatings are more likely to show interfacial failure, even if they are less well bonded than weak porous ones. Finally, increasing the friction between coating and scratch diamond will increase the interfacial traction and the crack lengths. More work is needed to quantify these effects and thus some caution is needed to compare the test results for very different coatings.

3.2 Relationship with Coating Performance

Initial results indicate that the ranking of coatings produced by a range of plasma spraying processes using the scratch test is similar to that which would be expected from component performance. For instance, the adhesion of air plasma sprayed oxide coatings onto a silicon carbide substrate was ranked using the test and both pure zirconium silicate and mullite coatings were found to show better adhesion than coatings of zirconia and alumina. This is due to the fact that they show a lower thermal expansion mismatch with the silicon carbide substrate. In thermal cycling tests the performance is correlated directly with the scratch test critical loads since the lifetime of the coatings was dictated by spallation (Fig. 11).

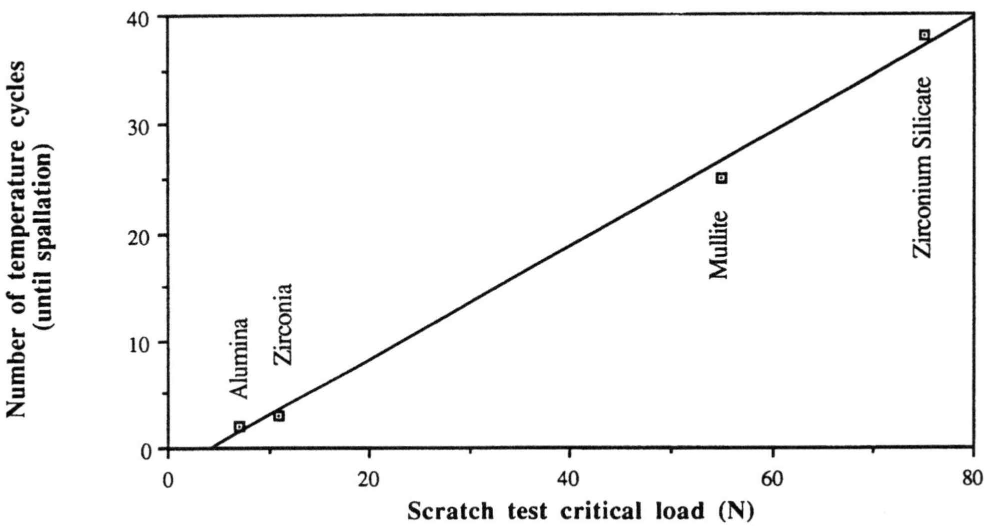

Fig. 11. Variation of the number of thermal cycles to failure with critical load for a range of oxide coatings plasma sprayed onto silicon carbide.

4. CONCLUSIONS

If carefully performed using a well-defined test procedure, the scratch test can give useful information about the adhesion of coatings. It is well suited to use as a process monitoring tool and it is sensitive to run-to-run variations in adhesion and other properties (substrate hardness and coating thickness). However, the correlation between component performance and scratch test results is not always good since component failure may be due to factors outside the test such as chemical reactions, substrate fracture or fatigue. Comparing the scratch test behaviour of different coating/substrate combinations is difficult unless some method of normalising for the intrinsic factors is used. Much more data are needed to establish a viable method for comparison.

ACKNOWLEDGEMENTS

This work was supported by the Corporate Research programme of AEA Technology.

REFERENCES

1. K. L. Mittal: *Electrocomp. Sci. Technol.*, 1976, **3**, 21.
2. A. J. Perry: *Thin Solid Films*, 1983, **107**, 167.
3. J. Valli: *J. Vac. Sci. Technol.*, 1968, **A4**, 3007.
4. P. A. Steinmann and H. E. Hintermann: *J. Vac. Sci. Technol.*, 1989, **A7**, 2267.
5. P. R. Chalker, S. J. Bull and D. S. Rickerby: *Mat. Sci. Eng.*, 1991, **A140**, 583.
6. S. J. Bull: *Advanced Techniques for Surface Engineering*, eds. W. Gissler and H. A. Jehn, Kluwer Academic Publishers, Dordrecht, 1992, 31–68.
7. P. J. Burnett and D. S. Rickerby: *Thin Solid Films*, 1987, **154**, 403.
8. S. J. Bull: *Surf. Coat. Technol.*, 1991, **50**, 25.
9. J. Seckler, P. A. Steinmann and H. E. Hintermann: *Surf. Coat. Technol.*, 1988, **36**, 519.
10. S. Yamamoto and H. Ichimura: *J. Mat. Res.*, 1992, 7, 2240.
11. S. J. Bull, D. S. Rickerby, A. Matthews, A. R. Pace and A. Leyland: *Proc. 1st Int. Conf. Plasma Surface Engineering*, E. Broszeit, W. D. Munz, H. Oechsner, K. T. Rie and G. K. Wolf, eds, DGM, 1989, 1227.
12. S. J. Bull and D. S. Rickerby: *Surf. Coat. Technol.*, 1990, **42**, 149.
13. S. J. Bull, P. R. Chalker, C. F. Ayres and D. S. Rickerby: *Mat. Sci. Eng.*, 1991, **A139**, 71.
14. W. D. Sproul: *Thin Solid Films*, 1983, **107**, 141.
15. J. Valli, V. Makela, A. Matthews and V. Murawa: *J. Vac. Sci. Technol.*, 1985, **A3**, 2411.
16. S. J. Bull, D. S. Rickerby, T. Robertson and A. Hendry: *Surf. Coat. Technol.*, 1988, **36**, 743.
17. S. J. Bull, D. S. Rickerby, A. Matthews, A. Leyland, A. R. Pace and J. Valli: *Surf. Coat. Technol.*, 1988, **36**, 503.
18. S. J. Bull, D. S. Rickerby and J. T. Gent: *Surf. Coat. Technol.*, 1991, 7, 145.
19. E. Lopez, F. Belzung and G. Zambelli: *J. Mat. Sci. Lett.*, 1989, **8**, 346.
20. R. Kingswell, D. S. Rickerby, S. J. Bull and K. T. Scott: *Thin Solid Films*, 1991, **198**, 139.

INDENTATION BASED MECHANICAL CHARACTERISATION OF THIN FILMS ON SUBSTRATES

M. V. SWAIN
CSIRO Division of Applied Physics, Lindfield NSW 2070
and
Department of Mechanical and Mechatronic Engineering, The University of Sydney, Sydney NSW 2006, Australia

ABSTRACT

This paper will discuss the ability of indentation methods to characterise the mechanical properties of thin films. The topics to be considered are: measurement of hardness and modulus from precision force–displacement data of pointed indentation tests; the use of spherical indenters to perform the same task and to determine a stress strain response; and the application of these indenters to measure film fracture and delamination.

1. INTRODUCTION

Indentation methods have been widely used for investigation of mechanical properties of materials for a long time. Very often, hardness is measured which provides information about the resistance of a material against penetration by another body. The measurement consists of pressing a hard indenter by a defined force into the test specimen and measuring the remnant imprint in the material after unloading. Usually, hardness is given as

$$H = P/A$$

The advantage of this kind of test is that it is a simple procedure with minimal specimen requirements (no special shape, one specimen can be used for several indentations, the specimen is not destroyed, only locally damaged). Moreover, the characteristic stressing during indentation is similar to many real situations (contact with other bodies, abrasion, wear, machining, etc.).

Originally, indentation methods were used to estimate the hardness of ductile, usually metallic materials. With technological development the field of their applicability has greatly increased. The need for a simple method of investigation of the mechanical properties of brittle materials led to the development of micro-hardness testers, with load ranges from 0.1 N to several N, and characteristic dimensions of the imprint given in μm^2. With these devices, it was appreciated that even such brittle materials as glass and ceramics may be plastically deformed in microscopically small volumes.

More recently, the need to study mechanical properties of surface engineered components has necessitated the development of so-called ultra microhardness testers, which use loads of only mN and are able to measure impressions only some nanometres deep. This situation is critical for thin films deposited on surfaces in order to achieve specific properties: mechanical, electrical or optical. Such advances have been possible due to improvements in precision measurement technology, and the use of computers for measurement control and data processing. In order to obtain as much information as possible, indenter load and displacement are measured during loading and unloading. This kind of measurement, producing P–δ (load–penetration) curves, enables one to determine not only hardness, but also the elastic modulus, stress–strain response (and how they depend on dimensions of the imprint). In this manner it is possible to reveal various processes and changes within the material, like pressure-induced phase transformations, initiation of cracks, etc.

The shortcoming of indentation methods occurs because of the strongly inhomogeneous stress field in the area of investigation. As a result, various methods for evaluation of results and determination of material characteristics have been developed and used from simple, often semi-empirical, to more sophisticated and complicated. The choice of a particular method depends on the material and property to be determined, and on the measuring techniques and system used. It also depends, among other things, on the form of the indenter. For example, pointed pyramidal indenters, like a Vickers or Berkovich, creates a permanent deformation in the specimen almost at the onset of loading, with complicated relationships between localised stresses and strains. Such indenters do have the advantage of geometrical similarity irrespective of the depth of penetration. On the other hand, a spherical-tipped indenter (with the tip radius of several μm) produces a purely elastic deformation initially and the force–displacement relationships for this case are relatively simple and well known. With such indenters geometrical similarity is not maintained during penetration.

The reliability of films or coatings not only depends upon their intrinsic mechanical properties but also upon the adhesion of the films to the substrate. In a recent review of film adhesion Mittal[1] listed over 300 approaches that have been developed to measure 'practical adhesion'. Two of the most popular being the 'pull-off' test of bonded stubs and scratch testing using spherical tipped Rockwell indenters (typically 200 μm radius). Both tests have serious limitations. Normal indentation tests with pointed or spherical tipped indenters also provide a basis to investigate this property.

In this paper the issues associated with the determination of the elastic and plastic properties using pointed and spherical indenters are outlined. This is followed by a critical appraisal of these concepts to the assessment of thin films. Finally, aspects of thin films fracture and delamination are discussed. In all areas the relevant equations are presented along with examples.

2. BACKGROUND

2.1 Pointed Indenters

The tacit assumption with such indenters is that they are perfectly sharp. The resultant force–displacement curve is as shown schematically in Fig. 1.

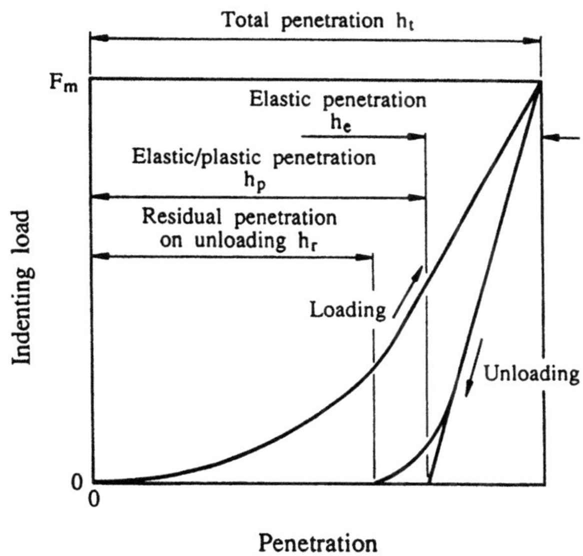

Fig. 1. Schematic force–displacement diagram for a pointed indenter and definition of various terms used in the text.

The displacement during loading consists of a number of components including: elastic displacement (h_e), plastic displacement (h_p), instrument compliance (h_c) and initial penetration (h_i) because with all instruments a finite force is required before the signal exceeds the background noise. That is the total displacement h_t may be written as

$$h_t = h_e + h_p + h_i + h_c. \tag{1}$$

A well calibrated instrument enables h_c to be substrated. An estimate of h_i may be obtained by extrapolating from the first 10–20 data points at low loads to predict the displacement at the load sensitivity limit of the instrument. However, as pointed indenters are not perfectly sharp but have a finite radius that may be 50 nm or greater, allowance must be made for the missing material otherwise the contact area will be overestimated as shown schematically in Fig. 2. Various approaches have been adopted to estimate the missing plastic displacement h_m. Doerner and Nix[2] used replicas to monitor the change of contact area with depth of penetration. Oliver and Pharr[3] estimated this correction assuming the elastic modulus of a material is constant as a function of depth of penetration. An alternative approach developed by Window *et al.*[4] used indentation of a hard metallic material and assumed the hardness remained constant. Such an approach enables an estimate of h_m as a function of depth which may be stored as a look-up table to correct the depth of penetration.

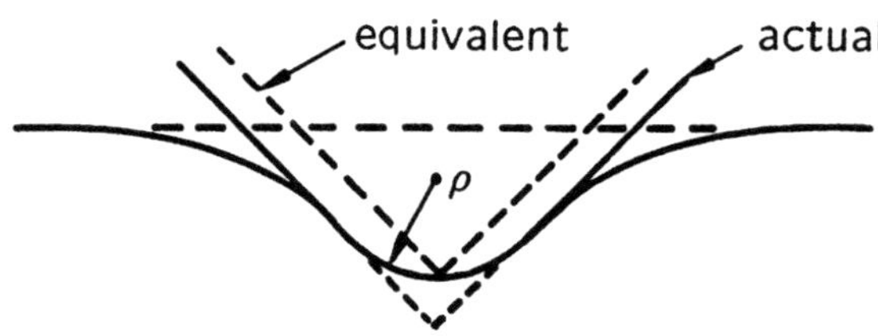

Fig. 2. Schematic drawing of actual versus ideal tip of a pointed indenter.

The elastic displacement h_e may be estimated from an approach proposed by Loubet *et al.*[5] based upon Sneddon's[6] analysis of a flat cylindrical indenter loaded onto a half space. The elastic recovery is then simply a linear fit to the unloading slope dP/dh at maximum load as shown in Fig. 1. Pharr *et al.*[7] have suggested this approach overestimates the elastic displacement because of the pointed nature of the indenter and they suggest that h'_e is $0.75\, h_e$. Thus one may now estimate hardness as:

$$H = P/A$$

where $A = k\, h^2_p$, and with $k = 24.5$ a constant for a Berkovich indenter, and

$$h_p = h_t + h_i + h_m - h_c - 0.75\, h_e.$$

The effective elastic modulus may be estimated from

$$E_e = \frac{dP}{dh}\frac{\sqrt{\pi}}{2}\frac{1}{\sqrt{A}} \tag{2}$$

where

$$\frac{1}{E_e} = \frac{1 - \nu_i^2}{E_i} + \frac{1 - \nu_s^2}{E_s}$$

where the subscripts i and s are for the indenter and substrate.

A major reservation of this approach at present is that it fails to consider 'pile-up' or 'sinking-in' about the indenter. Thus the value of the estimated h_p above may be considerably under or over-estimated because one of the assumptions present in the above approach is that the initial reference surface remains constant throughout the event. Another issue, as pointed out by Tabor,[8] is what does the estimated 'hardness' measure of a material that may work harden/soften, phase transform, crack etc.? To this end more recent definitions of hardness recommend the force–displacement curve be used to measure the work per unit volume of material displaced.[9]

2.2 Spherical Indenters

This approach, although used for over a century in its macro mode, and until recently only sporadically for micro- and nano-indentation,[10] has the ability to reveal substantially more about materials than pointed indenters. Spherical tipped indenters initially produce a purely elastic deformation resulting in the well known Hertzian relationship between force and displacement,

$$\delta = (9/16)^{1/3} (1/R_i)^{1/3} (P/E_e)^{2/3} \qquad (3)$$

where R_i = radius of the indenter. This relationship is valid only up to the moment when the maximum contact stress reaches the yield point, σ_y. At heavier loads the analysis becomes more complex,[10] but if one still assumes the unloading is elastic it is still relatively simple to estimate the contact pressure and elastic modulus. Figure 3 is

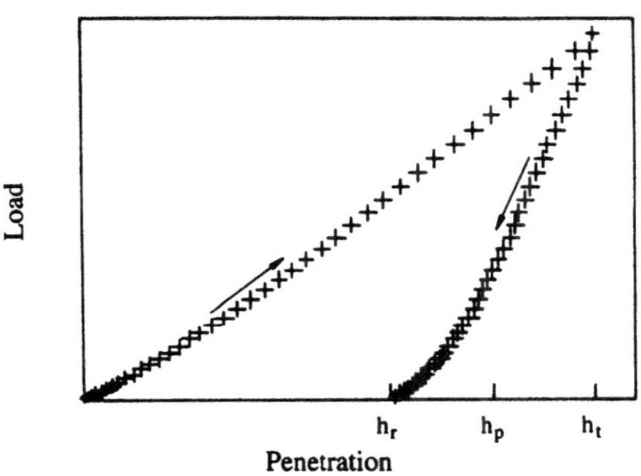

Fig. 3. Schematic of force–displacement data for a spherical tipped indenter.

a schematic figure of the load–displacement curve. The advantage over pointed indenters is that the plastic penetration depth h_p lies mid-way between h_t and h_r where h_r is the residual depth of impression. The contact pressure is then given by

$$P_m = P/\pi a^2$$

where

$$a = (2R_i h_p - h_p^2)^{1/2}$$

If the displacement is small compared to the radius of the indenter a simpler relationship between these parameters and the radii of contact is given by Johnson[11]

$$a^2 = \delta R_i.$$

An estimate of the elastic modulus may be obtained from the unloading curve by substituting $\delta = h_t - h_r$, and $^1/R_i$ by $(^1/R_i - {}^1/R'_r)$ into equation (3), where R_r is the residual radius of the impression given by

$$R'_r = (a^2 - h_r^2)/2h_r$$

(These terms/concepts are defined in Fig. 4.) Then the effective elastic modulus is given by

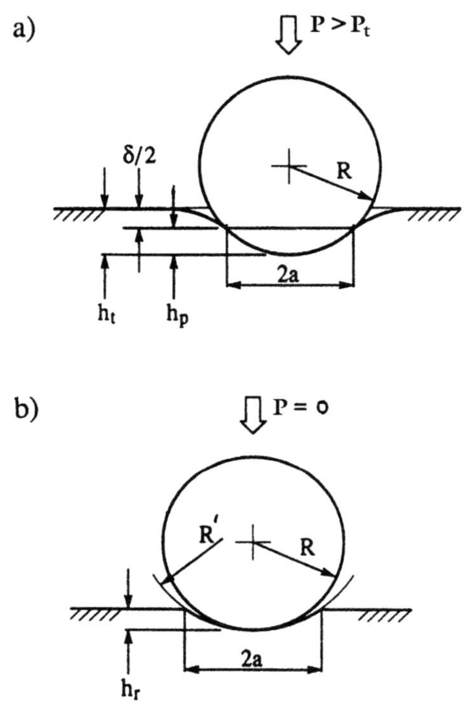

Fig. 4. Schematic of the contact of a spherical tipped indenter with a half space.

$$E_e = {}^3\!/_4 P/a\delta$$

The approach mentioned above suffers from all the issues mentioned with pointed indenters in that machine compliance, initial penetration depth and the indenter radius must be well defined. However, as is well known for spherical indenters the contact pressure and normalised contact diameter (to the indenter radius) may be used to construct a stress–strain curve, namely

$$P_m = 3\sigma_y$$

for a fully plastic impression (in a metal) and the strain is

$$\epsilon = 0.2\ a/R$$

Field and Swain[10] suggested a novel extension of this approach by indenting in a load–partial unloading manner as shown in Fig. 5. This enables h_r and h_p to be reliably estimated throughout the entire loading curve thereby following the contact pressure, modulus and stress–strain response of a material.

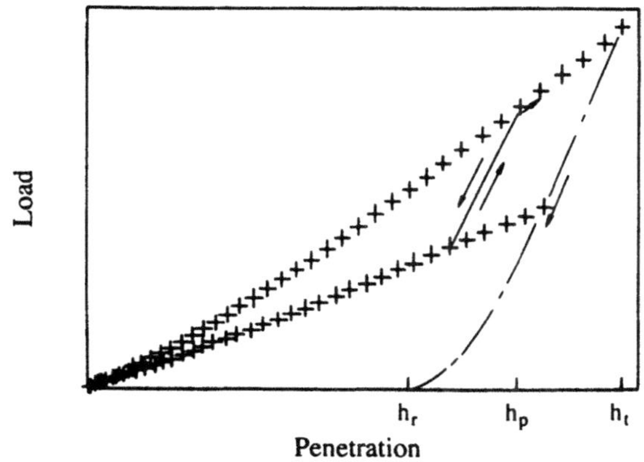

Fig. 5. Load–partial unload indenting procedure with spherical tipped indenter. Compare with Fig. 3.

The spherical indenter analysis outlined here suffers the same problems of 'pile-up' and 'sinking-in' as mentioned for pointed indenters. However, because spherical indenters transit from elastic to plastic behaviour they may be used to monitor the work hardening response. Field and Swain[10b] have used the recent theoretical analysis by Hill *et al.*[11] of the classic paper by Norbury and Samuels[12] of 'pile-up' or 'sinking-in' behaviour of metals to enable a correction of the data analysed in the manner outlined above.

2.3 Thin Films

A very important problem is the determination of properties of thin films that incorporates the influence of the substrate. The substrate with its properties (E, σ_y, H) influences the measured film behaviour, more particularly with thinner films. It is typically thought that as long as the depth of penetration is 10–20% of the film thickness substrate effects may be ignored. For deeper penetrations King[13,14] proposed an approximation for the elastic modulus with the introduction of an additional term in equation (2), namely

$$E_\epsilon = \kappa \frac{\mathrm{d}P}{\mathrm{d}h} \frac{1}{\sqrt{A}} \frac{2}{\sqrt{\pi}} \tag{4}$$

where

$$\kappa = \frac{(1 - \nu_f^2)}{E_f\,(1 - e^{-\alpha})} + \frac{(1 - \nu_s^2)}{E_s\,(1 - e^{-\alpha})} + \frac{(1 - \nu_i^2)}{E_i}$$

E_f, ν_f are the film modulus and Poisson's ratio and $\alpha = -\chi t/\sqrt{A}$ where χ is an empirical constant and t the film thickness.

An alternative approach has been proposed by Gao *et al.*[15] for films whose E values do not differ significantly from the substrate (~2), who considered the contact of a flat cylindrical indenter onto a film on a substrate. They proposed a simple asymptotic solution with the modulus and Poisson's ratio dependent upon the ratio of film thickness to contact radius. However, as pointed out by Yu *et al.*[16] there is little difference between the contact of a flat cylindrical and spherical indenter with a film on a half space.

The asymptomatic expression developed by Gao *et al.*[15] for the influence of a film on the elastic response as a function of the ratio of film thickness to contact radius (t/a) is given by

$$E_e = E_s + (E_f - E_s)\, I_o\,(t/a)$$

and

$$\nu_e = \nu_s + (\nu_f - \nu_s)\, I_l\,(t/a)$$

for the modulus and Poisson's ratio, with

$$I_o = \frac{2}{\pi}\tan^{-1}(\xi) + [(1 - 2\nu)\xi$$

$$\ln((1 + \xi^2)\,/\,\xi^2) - \xi\,/\,(1 + \xi^2)]\,/\,(2\pi(1 - \nu)]$$

where $\xi = t/a$ and the plot of I_o as a function of t/a is shown in Fig. 6.

Swain and Weppelmann[17] recently proposed that using this approach and indent-

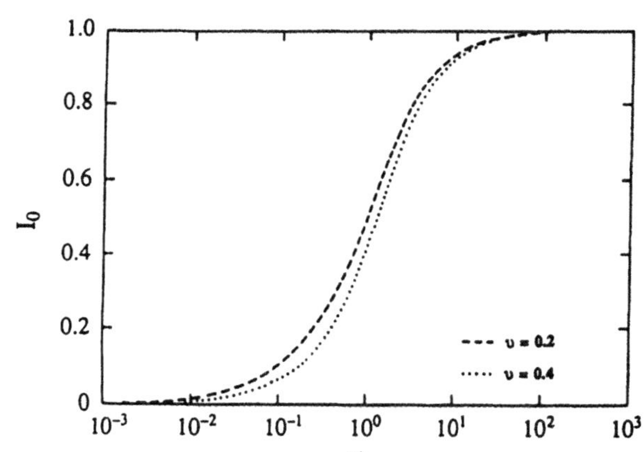

Fig. 6. Dependence of $I_o(t/a)$ with normalised contact dimension.

ing both the substrate with and without a film the modulus of the film could be estimated from the relationship

$$E_f = [E_e / E_s - (1 - I_o(t/a))] / I_o(t/a) \tag{5}$$

Generally, for the values of elastic modulus, hardness, etc. measured at very small depths of impression (compared with the film thickness), the influence of substrate properties is not so critical (Fig. 7).

Moreover, the character of the strain field, which is decisive for the total response of a structure (film-substrate), may be very different in the case of a ductile film on a hard substrate, or a hard, brittle film on a soft substrate (Fig. 8).

This complicates the evaluation of measurement. The general response, $P(\delta)$, of such a structure under local loading can be found quite easily, but the determination of particular properties of the film alone may be difficult. There are several approaches to the problem. For example, Battacharya and Nix,[18] starting from finite element elastic-plastic modelling of various film-substrate combinations, proposed the following expressions for the apparent (measured) hardness:

(a) soft film on a harder substrate

$$\frac{H}{H_s} = 1 + \left(\frac{H_f}{H_s} - 1\right) \exp\left[-\frac{\sigma_{Yf} / \sigma_{Ys}}{E_f / E_s}\left(\frac{\delta}{t}\right)^2\right] \tag{6}$$

(b) hard film on a softer substrate

$$\frac{H}{H_s} = 1 + \left(\frac{H_f}{H_s} - 1\right) \exp\left[\frac{-H_f / H}{\sigma_{Yf} / \sigma_{Ys}\sqrt{(E_f / E_s)}}\left(\frac{\delta}{t}\right)^2\right] \tag{7}$$

If we perform several measurements of hardness to various depths of indentation, we

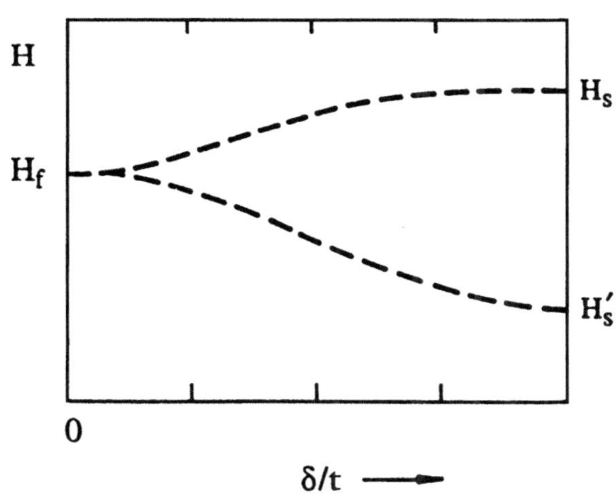

Fig. 7. Apparent hardness (H) of a film-substrate as a function of the relative depth of indentation (δ/t).

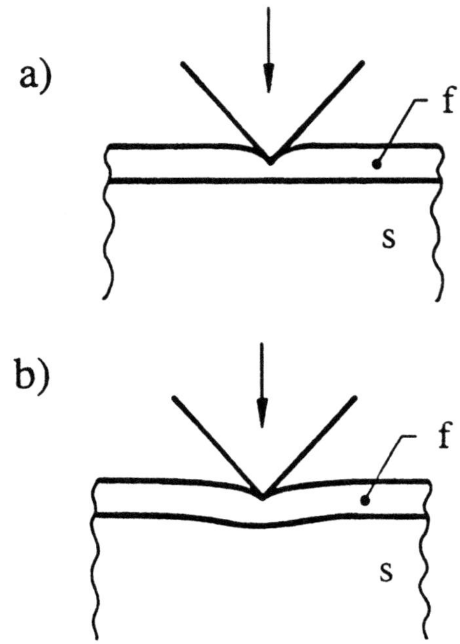

Fig. 8. Schematic illustration of (a) the behaviour of a soft film on a hard substrate, and (b) a hard film on a soft substrate.

can fit the measured H values by the functions (6) or (7) and, having known properties of the substrate, determine the hardness, elastic modulus, etc., corresponding to the film above. In the above-mentioned study,[18] starting from *a-priori* given values of E_f, E_s, σ_{Yf}, σ_{Ys}, . . ., the agreement between the observations generated by finite element modelling and those obtained using equations (6), (7) was quite good. With real measurement, however, especially when equations (6), (7) are used for prediction of properties outside the range of values used for their determination, the results may be less reliable.[19] The main reason is, perhaps, that the function of E_f/E_s and σ_{Yf}/σ_{Ys} in equations (6), (7) do not represent physical laws but are more a mathematical expression of observed trends.

As for hardness, there are many papers suggesting, in principle, that the apparent hardness of a film–substrate structure is determined by a weighted average of the volume of a plastically deformed material in the film (V_f) and in the substrate (V_s). The simplest formula,

$$H = \frac{H_f\ V_f + H_s\ V_s}{V_f + V_s}$$

originally proposed by Sargent,[20] was later modified in several ways to include the influence of further material characteristics and the extent of deformation developed; see, for example.[19] It has also been argued that not only the indentation depth but also the radius of contact area (together with the film thicknes) play an important role.

2.4 Film Fracture/Delamination

Thin films are often very brittle and deposited onto softer substrates to impart improved tribological, optical or optoelectronic functions. Indentation, particularly with pointed indenters, may often lead to film cracking and/or delamination. Analysis is often handicapped by the presence of very significant residual stresses in the films.

Limited attention has been given to the use of indentation to assess the adhesion of films onto substrates. Chiang *et al.*[21] used a Vickers indenter to estimate the interfacial toughness of ZnO films on glass which was later extended to include the effects of residual stress.[22] The relationship between interfacial crack size (see schematic drawing in Fig. 9) and other parameters was proposed to be

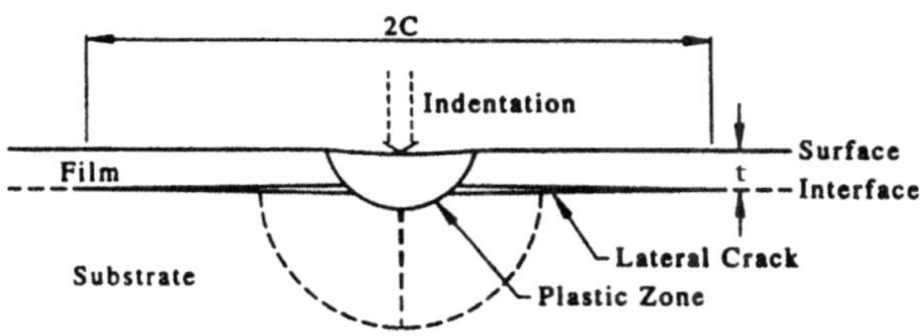

Fig. 9. Schematic of the interfacial cracking on which Chiang *et al.*[21] based their analysis.

$$c = \left[\frac{\gamma E}{t\,(1 - \nu^2)G}\right] \left(\frac{P}{H}\right)^{3/4}$$

where γ is a constant and G the interfacial toughness (ignoring the role of residual stress in the films). This analysis assumes that all the plastic deformation about the indenter occurs within the films and that the interfacial fracture is similar to lateral cracking.

Loubet *et al.*,[23] investigating brittle films on softer substrates, observed very minimal plastic deformation of the film beneath the indenter but rather extensive film cracking and substrate plastic deformation. They proposed a simple model relating depth of penetration to crack size, film modulus and thickness to interfacial toughness, namely

$$G = {}^{3}\!/_{8}\, E_c\, t^3\, d^2 \,/\, c^4 \tag{8}$$

where d is the depth of penetration at the edge of contact.

More recently Sugunuma *et al.*[24] have suggested that d in the above equation should be replaced by the remnant maximum uplift of the film at the contact site.

Rosenfeld *et al.*[25] have also presented an analysis of the problem of delamination of circular cracks about indentations of polymer films on rigid substrates.

3. OBSERVATIONS AND DISCUSSION

In this section examples of the response of various films loaded with pointed and spherical indenters will be presented.

3.1 Tantalum Oxide Films on Glass and Silicon

1 μm thick films were simultaneously deposited onto substrates of silicon and glass using ion-assisted (evaporation) deposition (IAD). These films are widely used in the optical industry as protective and selective transmission films.[26]

Observations of Berkovich indentations are shown in Fig. 10. Even at very low

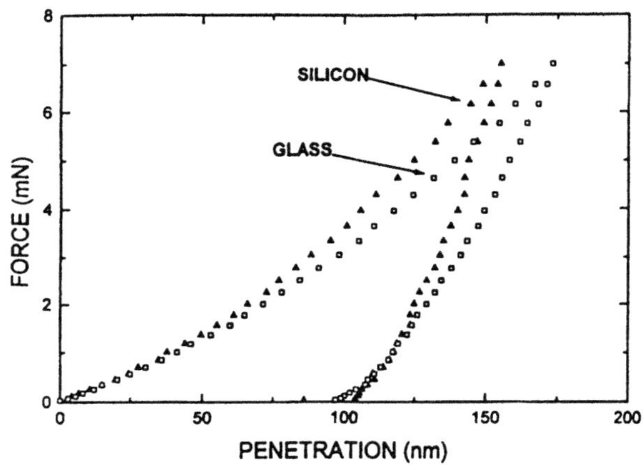

Fig. 10. Berkovich indentations of 1 μm thick tantalum oxide films on glass and silicon substrates.

loads (7 mN) with residual depths of penetration of ~1/10th of the film thickness, it is obvious that the lower modulus glass substrate (70 GPa) results in a greater displacement than for the stiffer silicon substrate (170 GPa). The calculated values for H were 7.2 and 9.5 GPa for the glass and silicon respectively. The situation becomes more pronounced at heavier loads.

Observations of 5 μm spherical indentation results are shown in Fig. 11. The data for films on glass and silicon are presented but the maximum depths of penetration on the silicon substrate were less. The hardness and modulus change with depth for both substrates as shown in Figs 11(a) and (b). In the case of the hardness results for the films on both surfaces it is noted that although they show a similar trend, the hardness on the silicon substrate is always higher than that on the glass. The modulus data, whilst also initially similar thereafter, diverge, reflecting the influence of the other substrate. No attempt has been made to incorporate the ideas of Gao *et al.*[15] in the reinterpretation of this data.

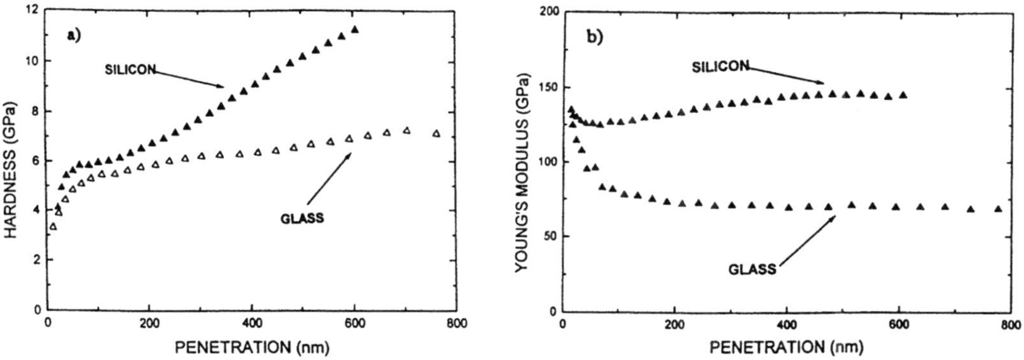

Fig. 11. Load partial unload analysed results for a 5 μm radius spherical indenter loaded onto 1 μm thick tantalum oxide films on glass and silicon. (a) Contact pressure, and (b) elastic modulus versus depth of penetration.

3.2 Al/Cu Films on Silicon

Al/Cu alloys are widely used in the microelectronics industry and bond pads and interconnects. They carry very high current densities and the fabrication and reliability of chips is often contingent on their mechanical properties, particularly adhesion.

Observations of pointed and spherical indentations on 1 μm thick films are shown in Figs 12 and 13. In both instances an inflexion of the force–displacement diagram

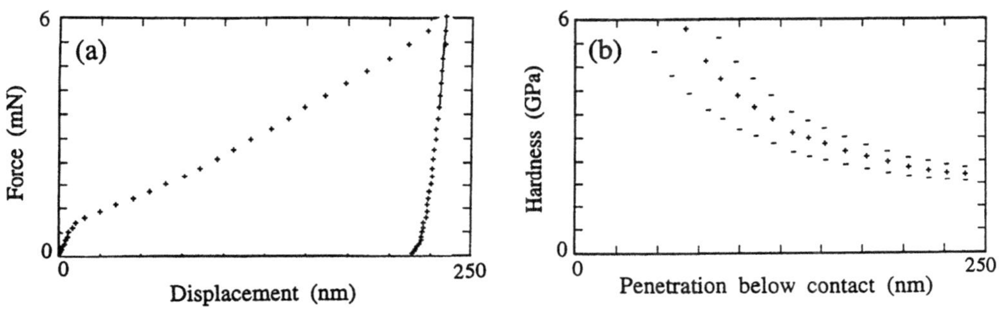

Fig. 12. (a) Berkovich indentations of a 1 μm Al/Cu thin film on silicon, and (b) variation of hardness with depth of penetration.

may be detected at very low loads. Interpretation of the hardness for the Berkovich data is shown in Fig. 12(b), and for the spherical data in Fig. 13(b). The spherical indenter clearly identifies an elastic response at very low load (the partial unload data superimpose on the loading points) before becoming elastic/plastic. The contact pressure versus depth plots show a very high stress to initiate plastic flow due perhaps to an oxide coating on the Al/Cu film. Similar behaviour has been observed for a number of metallic films.

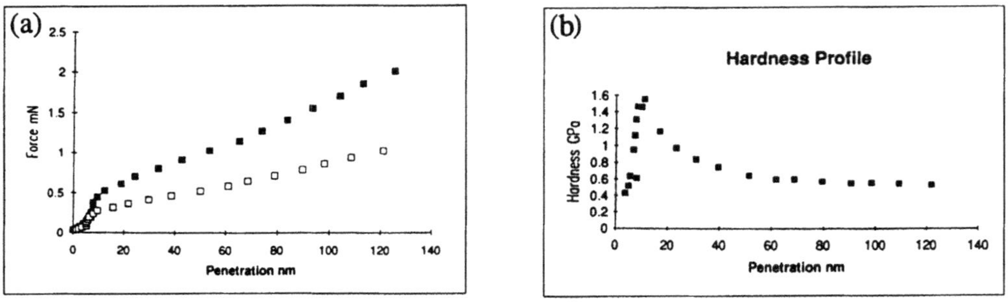

Fig. 13. (a) Load partial unload data for 5 μm radius spherical indentations of an Al/Cu film on silicon. (b) Mean contact pressure versus depth of impression.

3.3 Polyimide Films on Glass

Polyimide films are also widely used in microelectronics and where relatively severe conditions for a polymer prevail.

Observations of the force–displacement records for Berkovich observations on a 2 μm thick film deposited on glass are shown in Fig. 14(a). Analysis of this data is

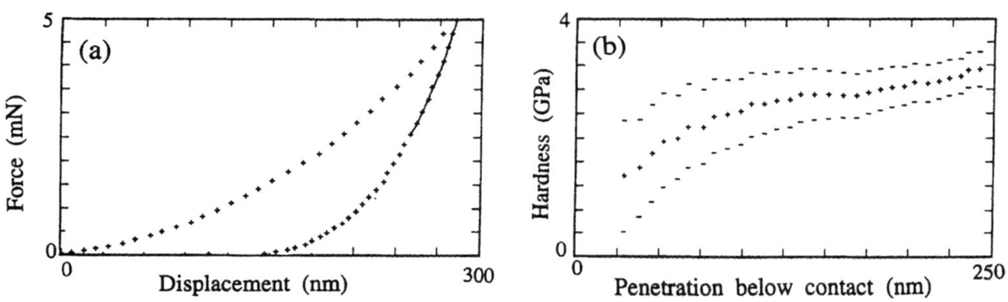

Fig. 14. (a) Force–displacement data for Berkovich indentations of a 2 μm polyimide film on glass. (b) Hardness versus depth.

shown in Fig. 14(b) and reveals that with increasing depth of penetration the hardness increases. The load partial-unload spherical indentation results are shown in Fig. 15(a) along with the evaluated hardness or mean contact pressure, Fig. 15(b), and modulus results, Fig. 15(c). The spherical indenter provides a clear indication of the change of elastic modulus with depth as would be anticipated from Fig. 6.

Another feature apparent at heavier loads and with both indenters was the inflexion of the force–displacement curve as shown in Fig. 16(a). The Berkovich indenter exhibits a slight discontinuity which was thought to be associated with delamination; however, for the larger spherical tipped indenters, Fig. 16(b), this feature was much more pronounced and highly reproducible.

It was also evident from optical microscopic inspection that a well defined delaminated circle of material remained about the contact site. This sharp dis-

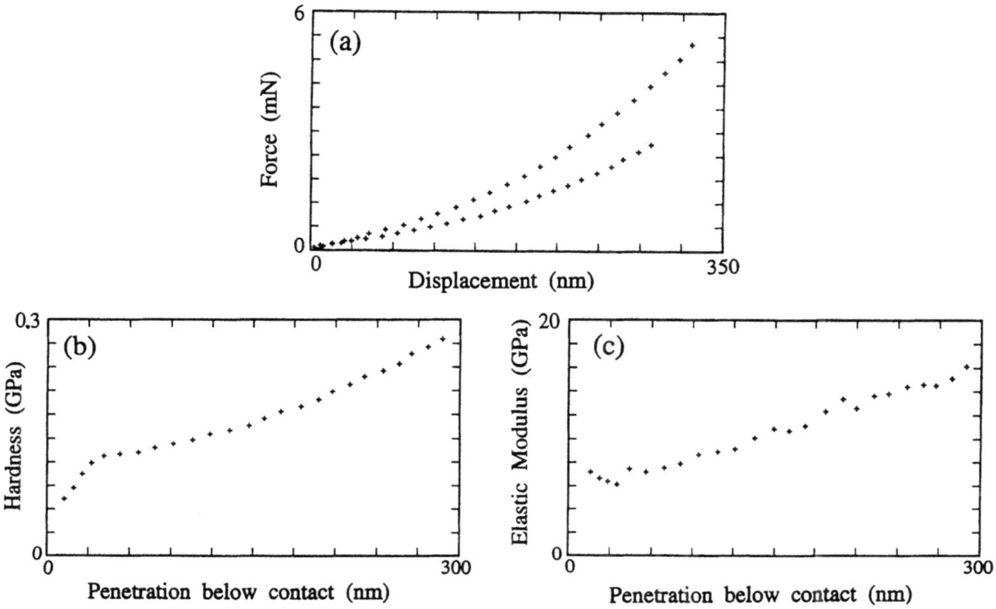

Fig. 15. (a) Load partial-unload data for indentations with a 5 μm spherical indenter on 2 μm thick polyimide films on glass. (b) Hardness and (c) elastic modulus versus depth of penetration.

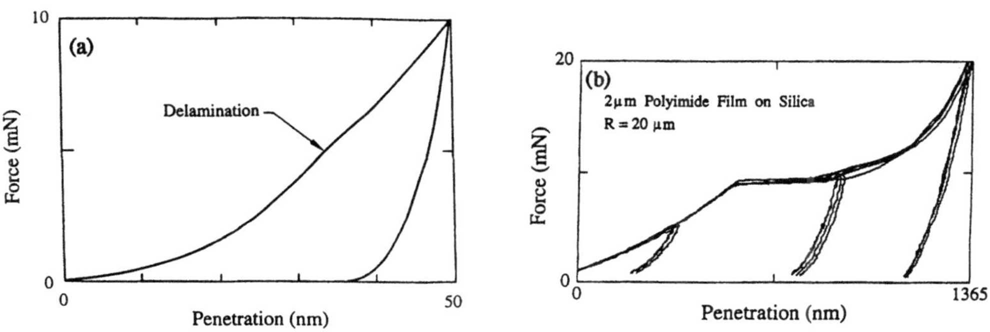

Fig. 16. Heavier load indentations of the 2 μm polyimide film on glass. (a) Berkovich, and (b) spherical indentation response associated with film debonding.

continuity in the case of the spherical indenter may be related to the critical interfacial shear stress as estimated by Matthewson.[27]

3.4 Ion Implanted Amorphised Silicon

Previous studies of the indentation response of silicon have been recently reviewed by Pharr[28] who pointed out the difficulties of interpretation of the force–displacement curves generated with a Berkovich indenter. Basically the many phase changes that occur during loading and unloading lead to hysteretic responses that jeopardise reliable interpretation of the data. However, Pharr did suggest from SEM observations of the residual impressions that the hardness of silicon slightly decreased with increasing indentation load, that is, it exhibited a slight indentation size effect (ISE).

More recently Weppelmann *et al.*[29] have investigated the indentation of silicon with small spherical tipped indenters and found that the complete loading and unloading response including hysteresis upon unloading could be interpreted and simulated to determine the phase transformation pressures. These authors concluded that the hardness of silicon was dependent upon the semi-conducting to metallic phase transformation (Si I > Si II) that occurs at ~11.5 GPa. In this section some observations of the influence of ion implantation to ~1 μm on the mechanical properties as recently reported by Williams *et al.*[30] will be described.

Figures 17(a–c) shows the force–displacement curves of crystalline silicon, ion implanted (amorphised silicon) and annealed amorphous silicon. The step during unloading is only observed for the crystalline silicon and the depth of penetration is least for this material and greatest for the amorphised material. Interpretation of data generated by the partial unloading procedure on all materials for the estimation of contact pressure is shown in Fig. 17(d). The initial response for all materials is elastic followed by, in the case of crystalline silicon, almost constant hardness whereas for

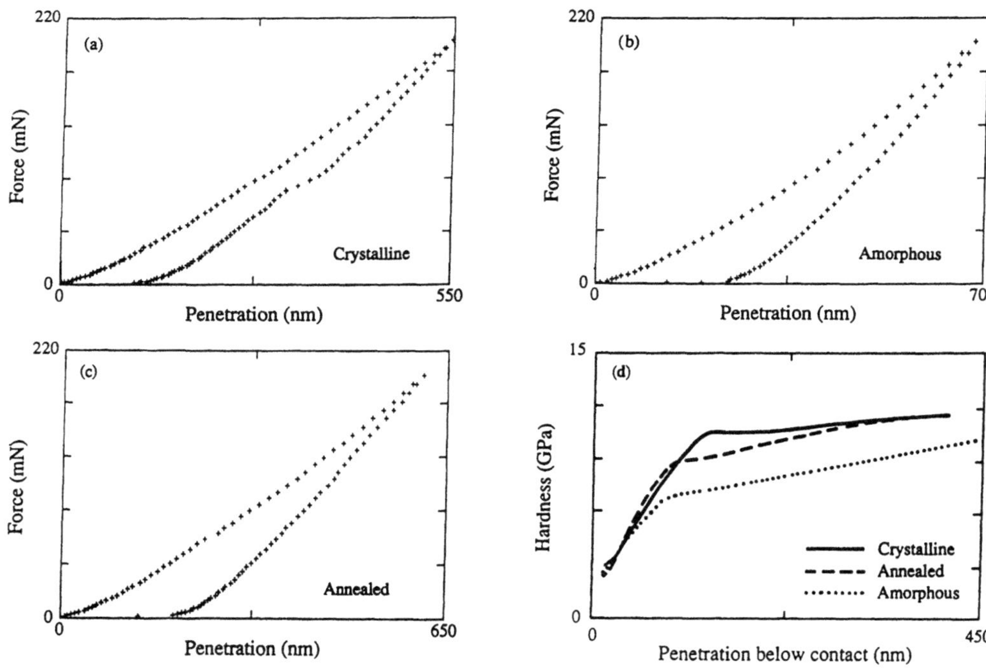

Fig. 17. Continuous loading force–displacement data with the 10 μm radius indenter for (a) crystalline, (b) amorphised and (c) annealed silicon. (d) Contact pressure versus depth of penetration for the three materials.

the amorphous material the yield pressure is much lower but increases with depth of penetration as the deeper crystalline material influences the data. The annealed material lies between these cases due presumably to some degree of lattice reconstruction.

3.5 SiO_2 Films on Steel

Observations of the force–displacement data of Berkovich indentations on a thin sol-gel SiO_2 film on steel are shown in Fig. 18. In all instances the data show a

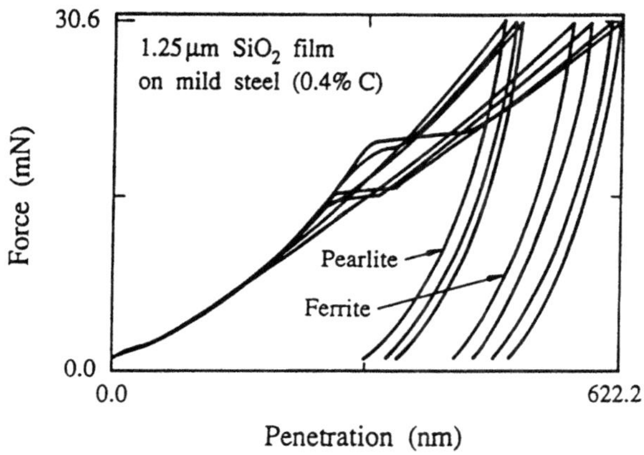

Fig. 18. Force–displacement data for a sol-gel SiO_2 film on mild steel with a Berkovich indenter.

discontinuity because of film cracking. Such cracking makes analysis for the determination of hardness problematical because of the unloading slope and therefore the modulus calculation and hence the plastic depth of penetration are determined by the steel's response. The variation of the breaking load for three film thicknesses was utilised to estimate the reliability of the film using Wiebull analysis[31] as shown in Fig. 19. The estimated values of the Weibull modulus (m) are listed in the figure and are typical of glassy brittle materials.

Another feature evident in the force–displacement curves is that the maximum resultant depth of penetration depends upon whether the indentation occurs above a pearlite or a ferrite grain of the low carbon steel.

3.6 TiN Films on Silicon

One final example is the force–displacement response of a brittle film on a substrate. TiN films are now widely applied to materials subjected to extreme tribological conditions such as drills, hobs etc. for metal cutting. Great emphasis has been placed upon the achievement of high hardness although the consequence of such is poor adhesion, as determined by scratch tests. The present observations are included in an endeavour to raise the question of what precisely 'hardness' does measure.

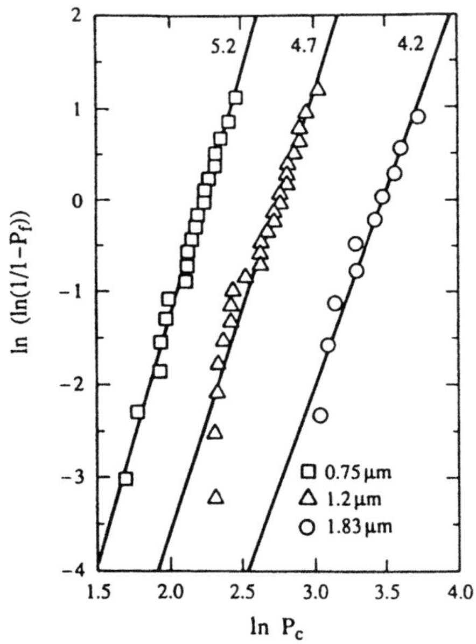

Fig. 19. Weibull modulus plots of the cracking load for three thicknesses of sol-gel SiO_2 films on mild steel.

Observations of the force–displacement data for spherical indentations of 2.7 μm thick TiN films on silicon deposited by the filtered arc method are shown in Fig. 20.[32] At lower loads the force–displacement data are entirely elastic.

It was such observations that Swain and Weppelmann[17] utilised to estimate the film modulus. At heavier loads, steps in the force–displacement response were noticed although upon viewing the residual impressions with high resolution SEM they still appeared as plastic impressions, Fig. 21(a). However, cross-sections of such impressions revealed a very different conclusion as shown in Fig. 21(b). The film

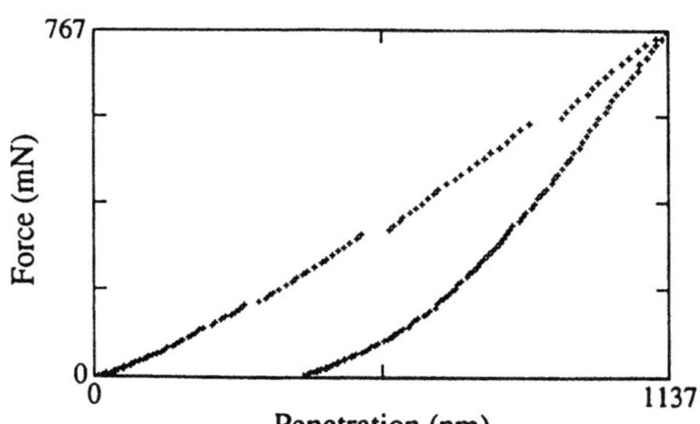

Fig. 20. Force–displacement data for a 10 μm radius spherical tipped indenter loading a 2.7 μm thick TiN film on silicon.

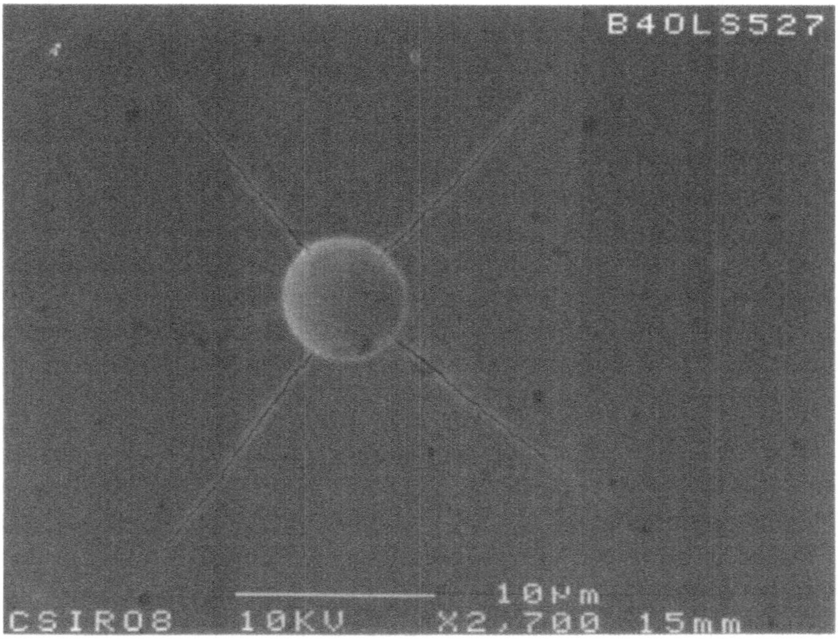

(a)

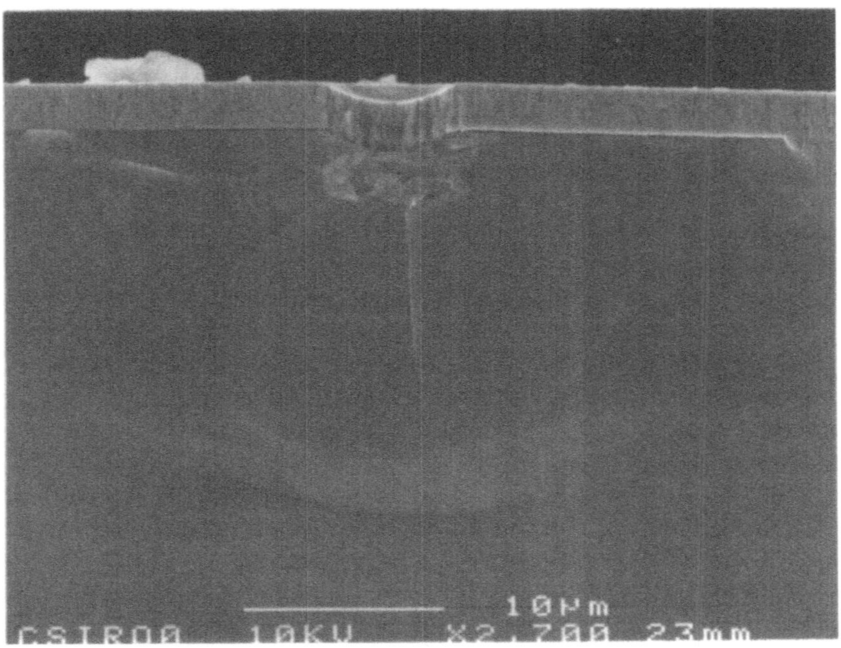

(b)

Fig. 21. (a) SEM observation of the remnant imprint of a 5 μm spherical indenter on a 2.7 μm thick TiN film on silicon at 500 mN. (b) Cross-section SEM of the same impression.[32]

exhibits a series of shear faults with virtually no evidence of plastic deformation. All the inelastic response is a consequence of the deformation of the silicon and various cracking about the impression. Other observations at lower loads reveal a similar behaviour as well as the presence of interlaminar cracks from which the interfacial toughness may be estimated.[32] With the aid of equation (8) the interfacial fracture toughness for the situation depicted in Fig. 21(b) was 0.1–0.2 $\mathrm{J m^{-2}}$.

The observations in Fig. 21 raise critical issues about what precisely is measured during indentation of highly elastic brittle thin films with Berkovich or spherical indenters. In the case of submicron thin films of these highly elastic materials such as diamond-like carbon (DLC), TiN, etc. it is the author's contention that deformation of the substrate precedes fracture of the films and that the interpretation of these data as a measure of hardness and modulus of the film is questionable.

4. CONCLUSIONS

Indentation with highly precise micromechanical probe systems provides a simple quantitative measure of the response of various films to contact loading. In this review the limitations of the interpretation of pointed and spherical indentation responses are outlined. There still remains many basic questions about the analysis of the contact response of thin films on substrates before satisfactory interpretation of the force–displacement data is possible. It is argued that spherical indenters provide a means of generating more useful data because they probe the elastic–plastic response to enable stress–strain behaviour far better than tests with pointed indenters.

It has also been shown that indentation may address issues of the reliability of films, particularly the magnitude and variation in fracture load and the delamination of films.

ACKNOWLEDGEMENTS

The author wishes to thank his many colleagues involved in the UMIS development at CSIRO and their collaboration on various projects which this paper has briefly reviewed.

REFERENCES

1. K. L. Mital: *Int. J. Adhesion Sci. Tech.*, 1987, **1**, 247.
2. M. F. Doerner and W. D. Nix: *J. Mater. Res.*, 1987, **1**, 601.
3. W. C. Oliver and G. M. Pharr: *J. Mater. Res.*, 1992, **7**, 1564.
4. B. Window, J. S. Field and T. J. Bell: *CSIRO Int. Rept.*, 1989.
5. J. L. Loubet, J. M. Georges, O. Marchesini and G. Nerille: *J. Tribology*, 1984, **106**, 43.

6. I. N. SNEDDON: *Int. J. Eng. Sci.*, 1965, **3**, 47.
7. G. M. PHARR, W. C. OLIVER and F. R. BROTZEN: *J. Mater. Res.*, 1992, 7, 613.
8. D. TABOR: 'Microindentation Techniques in Materials Science and Engineering', *ASTM–STP* 889, P. J. Blau and B. R. Lawn, eds., 1986, 129.
9. J. L. HENSCHELL, R. M. HOOPER, K. E. EASTERLING and W. B. LI: *J. Hard Materials*, 1993, **4**, 1.
10. J. S. FIELD and M. V. SWAIN: *J. Mater. Res.*, 1993, **8**, 297.
10b. J. S. FIELD and M. V. SWAIN, *J. Mater Res.*, 1995, **10**, 101.
11. K. L. JOHNSON: *Contact Mechanics*, Cambridge University Press, 1985.
12. R. J. HILL, B. STORAKERS and A. B. ZDUNER: *Proc. Roy. Soc. Lond.*, 1989, **A423**, 301.
13. R. B. KING: *Int. J. Solids Struct.*, 1987, **23**, 1657.
14. R. B. KING and T. O'SULLIVAN: *Int. J. Solids Struct.*, 1987, **23**, 581.
15. H. GAO, C-H CHIU AND J. LEE: *Int. J. Solids Struct.*, 1992, **29**, 2471.
16. H. Y. YU, S. C. SANDY and B. B. ROTH: *J. Mech. Phys. Sol.*, 1990, **38**, 745.
17. M. V. SWAIN and E. R. WEPPELMANN: *Proc. MRS Symp.*, 1993, **308**, 177.
18. A. K. BATTACHARYA AND W. D. NIX: *Int. J. Solids Struct.*, 1988, **24**, 1287.
19. B. D. FARBES, W. C. OLIVER, R. A. MCKEE and F. J. WALKER: *J. Mater. Res.*, 1992, 7, 3056.
20. P. M. SARGENT: 'Microindentation Techniques in Materials Science and Engineering', *ASTM–STP* 889, P. J. Blau and B. R. Lawn, eds., 1986, 160.
21. S. S. CHIANG, D. B. MARSHALL and A. G. EVANS: *Surfaces & Interfaces in Ceramic and Ceramic Metal Systems*, J. A. Pask and A. G. Evans, eds., Plenum Press, N.Y., 1981, 603.
22. D. B. MARSHALL and A. G. EVANS: *J. Appl. Phys.*, 1984, **58**, 2632.
23. J. L. LOUBERT, J. M. GEORGES and P. L. KAPSA: 'Mechanics of Coatings', *Proc. Leeds–Lyon Symp. on Tribology*, 1989, Elsevier, 1990, Paper XIX, 429.
24. M. SUGANUMA, M. V. SWAIN and E. R. WEPPELMANN: in preparation.
25. L. G. ROSENFELD, J. E. RITTER, T. J. LANDNER and M. R. LIN: *J. Appl. Phys.*, 1990, **67**, 3291.
26. P. J. MARTIN, A. BENDAVID, M. V. SWAIN and R. P. NETTERFIELD: *Proc. MRS Symp.*, 1993, **308**, 583.
27. M. J. MATTHEWSON: *Appl. Phys. Lett.*, 1986, **49**, 1427.
28. G. M. PHARR: 'Thin Film: Stresses and Mechanical Properties III', *Mat. Res. Soc. Proc.*, 1993, **239**, 301.
29. E. R. WEPPELMANN, J. S. FIELD and M. V. SWAIN: *J. Mater. Res.*, 1993, **8**, 830.
30. J. S. WILLIAMS, J. S. FIELD and M. V. SWAIN: 'Thin Films: Stresses and Mechanical Properties IV', *Mat. Res. Soc. Proc.*, 1993, **308**, 571.
31. M. SAMANDI, T. J. BELL and M. V. SWAIN: 'Thin Films: Stresses and Mechanical Properties II', *Proc. Mat. Res. Soc. Symp.*, 1992, **239**, 355.
32. E. R. WEPPELMANN, X. Z. HU and M. V. SWAIN: *Int. J. Adhesion Sci. & Tech.*, 1994, **8**, 611.

SOME CURRENT QUALITY ISSUES IN ADVANCED SURFACE ENGINEERING: A CONSPECTUS

K. N. STRAFFORD*
Ian Wark Research Institute, The University of South Australia, SA 5095

1. INTRODUCTION

A central theme of this book is the need for all persons concerned with the field of Surface Engineering (SE) and especially its advancement – at the *research and development* (R & D) level, at the *coating supplier* level, and, especially, in the context of the *user* (i.e. the agent who, ultimately, must be concerned with the *commercial success* of, for example, an advanced coating system which critically depends on the achievement and delivery of quality-for-performance) – to be aware of the current difficulties and problems which have to be addressed in the broad area of quality.

At the R & D level, scientists and engineers need to acquire systematically a much better basic knowledge of the various process technologies, e.g. in the PVD processes plasma densities and their significances to permit the repeatable creation of novel advanced SE systems with outstanding properties and performances, with an obviously potentially competitive edge on older systems produced by more conventional treatments/processes. This task, in fact, is a daunting one, and an enormous challenge.

Also at the fundamental level there is a need to understand structure/property relationships in, for example, coatings, so that, ideally, such entities could be designed, *ab inito*, to develop or display desired properties. At the present time such understanding does not, in general, exist, a situation which is in marked contrast to the progress made in the past 20–30 years or so in the fields of materials science and engineering in relation to the creation of bulk, monolithic engineering materials. In this context, perhaps, the micostructural design and associated sophisticated production processes to create single crystal gas turbine blades of complex composition via investment casting and thermal treatments to yield components with precise adequate creep life and the necessary resistance to environmental degradation (oxidation and hot corrosion) is a useful example.[1] Paradoxically, the scope for the

* Present address: The Institute of Materials, London SW1Y 5DB, UK.

creation of truly tailored coatings of chosen composition and structure (and therefore properties), including multilayer format, via such flexible technologies as PVD and CVD, in their many variants, is considerable.[2] Ultimately it should become possible to design and select a coating or coatings system (recognising the important role and support of the associated, possibly engineered, substrate) via an expert system, thereby achieving true fitness-for-purpose.

Also at this R & D level a further most critical issue is the precise significance of measured properties and characterisation, e.g. hardness, adhesion, porosity, thickness etc.[3] in relation to actual coating performance. Thus hardness is a commonly quoted parameter and is obviously some measure of the ability of a coating to resist wear. What is needed, however, is an awareness of the significance of particular hardness levels (and other properties) in relation to wear performance, so that a coating engineered to a particular hardness or hardness range could be reasonably expected to offer understood and anticipated tribological behaviour: that is, a particular wear life could be *designed* into a coated component. A much beter corresponding knowledge and understanding exists in the world of hardened or hard monolithic components, brought about by particular chosen mechanical and thermal treatments, which, however, are often well-defined in exising quality assurance/control standards.

It is believed that considerable progress could be made in this area of the philosophy and rationale of the design of coatings by analysing the performance of 'proven' coating systems, for example, in multilayer coated tools. Thus the role of individual layers of material in a multilayer coating in resisting wear and failure at particular critical geometrical parts of the tool needs to be systematically examined. Such a systematic forensic approach should allow extrapolation into advanced coating design with more reproducible enhanced, accountable, properties. Strafford and co-workers have discussed this further.[4–6]

The *coating producer* and *supplier*, in effect, are the central players in the creation and development of a commercial coating through a suitable process technology which often will have been scaled-up from research and development findings. The company will need to be sufficiently in tune with current R & D in the specific coatings area in question, and, in particular, be aware of the significance of the properties and characteristics of coatings in relation to anticipated performance demand by the coating user – the purchaser and commercial customer. Clearly it is fundamental that, especially in the creation of a new novel coating, the organisation also understands something of the fundamental aspects of the design of the coating equipment, and especially how the process variables may be manipulated and controlled to permit the repeatable deposition of material(s) with appropriate characteristics/properties and therefore performance. Often such details will be determined by the equipment supplier, the coater operating an essentially turn-key system. Even here, however, some minimal level of undertaking of the basics of the deposition technology, e.g. PVD, will be mandatory.

The *coater* is, of course, aware of the role and importance of other variables, e.g. component surface finish and condition, and size and geometry, on the efficacy of the deposition process and especially some of the basic qualities of the coatings, e.g. porosity, coherance and adherence to the substrate, essential to a successfully-created coating. In this context, the coating supplier must be in a position to offer advice, as necessary, to the coating user. This, ideally, would involve the supplier in the earliest stages of the design process for the engineering artefact. It follows that the knowledge and know-how of the supplier will need to be incorporated directly or indirectly in the specification for the component and, more generally, be drawn into the discussion process when quality assurance and control guidelines are being developed and drawn up to create a definitive standard for the production of a new, novel or advanced coating.

Finally, in this tripartite relationship concerning recognition of the need for, and the creation and the implementation of, quality standards, attention can be directed at the role of the *coatings user*. Truly it can be said in this context that 'the proof of the pudding is in the eating'. The coatings' user (commonly a manufacturer) is ultimatley responsible for quality in performance of the coated entity to the satisfaction of the component user, i.e. the customer. In the drawing up of specifications and in the coating design process itself (at the R & D level referred to earlier), the user obviously preferably should be involved. While it is probably true to note that with, for example, conventional electroplating processes, where in general adequate quality standards exist in relation to attributes and function, this scene is most certainly not true in the field of advanced coatings. While it is becoming more and more common for the manufacturer, e.g. an automotive 'manufacturer', to insist on the delivery of various parts, components or elements *to agreed standards* (and the manufacturer in fact is seen rather as an assembler of parts), the manufacturer cannot, ultimately, be ignorant of the difficulties and problems which often beset the supplier of the coated component, entities which may be partially removed via the creation of standards. In this process the coatings user must be proactive.

In summary, the availability of appropriate quality assurance and control standards for advanced coatings is crucial. In the rapidly-developing field of advanced coatings produced by the newer process technologies, where there is, in general, a dearth of understanding at the fundamental level, both in respect of coating structure/property relationships and process and machine variables, and in coating performance, there is an urgent need for the three parties to recognise the difficulties and cooperate with Standards Organisations to create a user-friendly standard and guide to ensure the successful commercial development of the many promising 'experimental' newer coating systems which are emerging and have the potential to greatly enhance manufacturing efficiency.

2. SOME ISSUES AND PROBLEMS

As noted earlier, for sometime and especially over the past 3–4 years – to judge, *inter alia*, from the emerging open literature, including that derived from/associated with SE Conferences and Workshops worldwide – there has been increasing concern regarding quality assurance and control issues in association with the advanced technologies.[7–9] At the same time there is also continuing interest in the design and development of an automatic, expert-system, control of deposition processes, a topic which is central to quality issues.

In attempting to identify and analyse some of the more important issues of current (and no doubt) continuing concern it is useful to focus on three areas: (1) deposition processes and process control issue; (ii) characterisation and intrinsic coatings' properties; and (iii) coating testing and performance evaluation. Many of the points to be raised within these headings have been the subject of continuing discussions in the literature, and indeed were a focus within a live Forum Discussion as part of the Workshop held in Adelaide in 1994, the papers from which have formed the basis of the text of this book.

Under (i), it is well recognised that in, for example, PVD, a large variety of machine and environment parameters are involved in the creation of a chosen thickness of a given coating, e.g. TiN of controlled properties – especially hardness, stoichiometry, colour, adhesion, bias voltage, argon and nitrogen gas pressures, substrate temperature.[10] The critical importance of process variables has also been demonstrated in the deposition of metallic coatings.[11] While from the materials science point of view it would be reasonable to expect such and other measurable properties/characteristics to be critically important with respect to coating performance, e.g. wear, the actual importance or even relative significance of which has yet to be established in definitive experiments. Ultimately, to be able to design-in chosen properties for particular service conditions, a deep understanding of such matters will be required, albeit some parameters being more critical than others. Strafford and co-workers,[3,12] Bull,[13] Saunders and Bull[7] and Matthews and Leyland[14] have discussed these matters. From a practical point of view (in terms of machine/process control and ease of acceptance etc.), a key question is what is the *minimum* number of easily measured/controllable parameters necessary to define reproducible performance?

It is well recognised that hardness is a good indicator, at least, as to the wear potential of a coating. Fundamentally, however, there is a need to define hardness *at this level* for thin or very thin coatings. The potential of the nanoindentation technique in deriving a number of parameters such as elastic modulus and recovery characteristics is challenging.[15] Another interesting area here is the potential of microhardness measurements to ascertain fracture toughness, drawing analogies with the world of monolithic ceramics.[16]

A further aspect of significant theoretical and practical importance is the surface finish/condition of the artefact to be coated. Subramanian and co-workers,[4] for example, have demonstrated how surface finish, i.e. physical roughness affects coating adhesion and hence tool performance.[17] Certainly the physical detail of surface roughness is followed by thin TiN coatings:[4] that is *surface condition* is important in the drive to create sound reliable coating systems. At the same time the very real cost of creating unnecessarily fine-smooth-surface finish in the hope of obtaining better adhesion and performance has to be recognised.[18] Often heat-treated tools and dies may have contaminated surfaces with partial oxide or oil coverage and, in any case, a thorough clean/degreasing treatment prior to coating must be a top priority.

Again, the level of residual (internal) stress in coatings is known to be a significant variable influencing coating integrity and wear behaviour. It must be assumed (by analogy with electrodeposited coating systems,[19]) that a degree of internal stress will be inherited from the substrate.

More generally, there is evidently a need to examine the applicability and approach of existing standards, e.g. AS3902, to thin film technology. As always, there is an urgent need for technology transfer – the provision of adequate education and training in these issues, a point which has been stressed elswhere by Strafford *et al.*[20]

3. CHARACTERISATION AND INTRINSIC COATING PROPERTIES

It was noted earlier in Section 1, within the context of the pursuit and achievement of quality via appropriate R & D effort, that a basic issue relates to the difficulties in the actual characterisation of the nature (physical attributes and chemical composition and structure) of thin coatings, and the determination of many of their properties, especially mechanical properties. Strafford and co-workers[3] have provided and discussed a comprehensive list of characteristics and properties which, from a fundamental point of view, are all necessary to fully define the coating. This includes: density, composition and stoichiometry, structure, morphololgy and microstrucure, surface roughness/finish, thickness, adhesion, cohesion, hardness, internal stress, strength, fracture toughness, ductility, elastic modulus, friction coefficient, thermal expansion coefficient and thermal conductivity. Further, it has been pointed out[3] that, in practice, there is a conflict or at least a distinction between the characteristics/properties of coatings as measured/monitored by materials experts (the scientist) and those needed for design purposes by engineers – a 'communication gap'.

Thus it has been suggested that in the first category are composition, thickness, hardness, adhesion, friction coefficient, and wear rate, while parameters such as Young's Modulus, Poisson's ratio, coefficient of thermal expansion, thermal conductivity, density, specific heat and stress-temperature behaviour fall into the second class.

This obviously cannot be a 'water-tight' designation, but such categorisation is useful for analytical purposes when attempting to decide key parameters which will undoubtedly play a major role in attaining and maintaining coating integrity and performance, and hence are likely to be of major significance in a quality audit – without control/assurance of these properties reproducibility in performance, and ultimately performance *optimisation* cannot be achieved.

There is also a need, however, to recognise that a 'priority' list of evidently-critical coating parameters can be drawn up in the context of tribological performance, as suggested by Matthews *et al.*[14] and by Saunders and Bull.[7] This list includes just seven properties/characteristics – thickness, hardness, adhesion, friction, wear, composition and residual stress. For the purpose of discussion, perhaps these could be coded as 'engineering' requirements. All these authors believed this portfolio of properties could adequately define performance and could, in general, be readily measured, with the possible exception of residual stress. The significance and the problems of assessing internal stress are discussed in this text.[21]

Finally, taking a more pragmatic viewpoint, Matthews and Leyland,[14] quoting the results of a survey among coaters and users in the UK, identified ten properties with which there was a general familiarity and feeling as to importance – adhesion, thickness, surface finish, corrosion resistance, friction and wear performance, hardness, porosity, composition, internal stress and structure/morphology.

It is evident therefore that there are a large number of intrinsic properties and characteristics which, ideally, should be available for the use of the coating designer – a comprehensive data base analogous to thermodynamic property data bases. While in the practical/engineering context this list can be reduced in a rational way, in some cases there are real difficulties in actually attempting to make measurements/determine properties, principally because the coating or film is very thin. For example, chemical/structural characterisation by X-rays can be problematical because of the interference from the substrate. Again, with hardness measurements there may be a contribution to the determined value from the substrate. This may partially account for the extraordinary scatter in hardness values reported in the literature for ceramic films, in contrast to bulk ceramics where less variability is noted.[22]

Also in the context of hardness values and determinations for thin films, the very high values reported for some compounds by certain authors is of interest. Strafford *et al.*,[23] for example, have noted the very major influence deposition parameters, especially bias voltage, in PVD processing have on hardness, apparently principally reflecting measurable changes in stoichiometry in compounds generic to TiN – ternary and especially quartenary compounds such as TiVN, TiAlN, TiHfN and TiNbN, and TiAlVN, respectively. These coatings are currently of considerable interest as the next generation of coatings for tools and dies etc. It also seems likely that the level of internal stress in hard coatings will influence hardness, and this may also partially account for the very high values reported.

Because of the thinness of many of these coating systems, particularly those in multilayer format,[24] there would seem to be a real potential for, and a need to establish more thoroughly, the role/purpose of nanoindenation methods.[15,25] An additional attraction here is the simultaneous ability with this technique to determine elastic and plastic (relaxation) characteristics of thin films. However, the whole basis of the technique and its role in characterising engineering coatings, including its relationship to and overlap with more conventional and better-established microhardness measurements, needs systematic attention.

Because of the high hardness (and possibly brittleness) of thin ceramic coatings it may be possible to measure fracture toughness via indentation methods, following the established philosophy for bulk monolithic ceramics.[16] In fact, of the various mechanical and physical properties listed earlier, the general absence of fracture toughness data for ceramic films and coatings is most unfortunate. A knowledge of this parameter and especially the manner in which it might systematically improve by *microstructural engineering* – e.g. by precipitation and particle reinforcement – to create improved tribological performance would be immensely beneficial for coating design purposes.

4. TESTING AND PERFORMANCE EVALUATION

The tenuous and poorly understood links between coatings' propeties and characteristics, and actual tribological (or corrosion) performance – compatability/interaction with the work environment – have been stressed earlier in Section 3. It is sufficient to reiterate that this area is significantly problematical in the pursuit of quality-fitness for purpose – especially in the identification of critical parameters for the establishment and drawing up of accepted standards.

In any area of testing, however, there is often the need to be able to conduct tests/evaluations in an accelerated mode (usually in the laboratory) and this presents new problems. In the area of tribology, for example, there are numerous recognised methods of testing to arrive, at the least, with a relative ranking of materials as to wear performance. Furthermore, in the laboratory, it then becomes possible to systematically vary perceived significant variables, e.g. stress level, temperature, working environment etc., to model wear behaviour, and hence to understand wear mechanisms. Such a forensic approach, obtaining an appreciation of the approach to failure in say a coated tool, either for single point turning or in drilling is clearly necessary to begin to confidently create and offer design information – the most appropriate coating formulation and selection.

For example, recent studies conducted by Strafford and co-workers[5,6] have identified wear modes and mechanisms in such tools, where a systematic comparison of uncoated tools with their various coated counterparts has been made. In particular, the wear patterns experienced with a number of multilayer coating systems have been examined, e.g. carbide tools coated with Ti(C, NN)/TiN or $TiC/Al_2O_3/TiN$. This analysis has allowed identification of the roles played by these various coating layers on a number of critical tool areas engaging with the workpiece. A key aspect in these studies has been the measurement and linking of cutting/drilling forces in single point turning and drilling, and the associated areas of wear of the tools leading to the onset of failure.[26,27]

Such controlled laboratory studies undoubtedly yield much useful and interesting information of a direct relevance to tooling design, and therefore quality in performance. However, it has been recognised that such tests tend to be situation-specific, and there is a need in general to establish the validity and correlation between laboratory tests and actual field testing. Part of this approach will be to conduct similar forensic analysis on coated components in the field, to ensure that at least broadly similar failure mechanisms are operative.

Within the context of establishing quality in performance standards, then there is a need to agree and devise appropriate testing methods for, say, coated tools and dies. Some standards for uncoated tools are already in existence (e.g. ISO 3685), but there is a dearth of methodology and tests for advanced surface coated systems. In the establishment of such information, there is a need for national and international cooperation to conduct 'round-Robin' tests to be able to understand and attach levels of importance to particular details. Some progress in this area has already been made within the European Community.[28]

5. THE WAY FORWARD

It is evident that the issue of quality assurance and control has yet hardly been addressed in the field of advanced tribological surface coatings. This is against the background of the enormous advances made to the associated processing technologies – CVD and especially PVD – over the past decade or so. Again, drawing largely from the related but different monolithic world of bulk ceramics (albeit justified with some encouraging and promising performance results from some novel experimental coatings, e.g. Al_2O_3, ZrC, TiAlN, TiAlVN, etc.), it would appear that a number of new coatings materials, and especially coatings sytems in multilayer format, have real potential.

However, even with existing systems such as TiN there is some concern about the lack of reproducibility in performance: certainly, there must be concern as to whether the system (tool plus coating) has been optimised.

It has been argued that the quality and the successful development acceptance of new novel advanced systems will depend critically on: (i) in-depth fundamental studies and understanding of the basic property of coatings and their origins i.e. structure/property relationships; (ii) a better understanding of process equipment and variables, including any scale-up problems to be able to tailor make and yield new coating systems, in accordance with the design factors based on knowledge from (i); (iii) confidence in instruments and equipment to be able to monitor/measure coating properties/characteristics; and (iv) a better understanding of the significance and importance of such properties/characteristics in relation to desired tribological/corrosion performance.

A major present difficulty is the largely uncoordinated R & D effort in the industrialised nations in 'the field' in these areas of processing and equipment, characterisation and especially performance testing. Collaboration between the research scientist and engineer, the coater and the user is needed to provide a forum for discussion, with a view to creating acceptable procedures and standards. Government agencies and laboratories, academic establishments and leading manufacturers, worldwide, need to actively collaborate in order to develop quality standards in the field of advanced surface coatings and provide confidence if the full potential of this technology is to be realised.

REFERENCES

1. K. N. Strafford: *The Corrosion Behaviour of High Temperature Alloys in Gaseous Environments, in High Temperature Alloys – Their Exploitable Potential*, J. B. Marriott, M. Merz, J. Nihoul and J. Ward, eds, Elsevier, 1987, 53.

2. A. Matthews: *Surface Engineering Practice – Processes, Fundamentals and Applications in Corrosion and Wear*, K. N. Strafford, P. K. Datta and J. S. Gray, eds, Ellis Horwood, Chichester, UK, 1990, 33.
3. K. N. Strafford, C. Subramanian and T. P. Wilks: *J. Materials Processing Technology*, 1993, **38**, 431–448.
4. C. Subramanian, K. N. Strafford, T. P. Wilks, L. P. Ward and W. McMillan: *Surface and Coatings Technology*, 1993, **62**, 529.
5. C. Subramanian, K. N. Strafford, T. P. Wilks, L. P. Ward and M. McPhee: *Surface and Coatings Technology*, 1993, **62**, 641.
6. K. N. Strafford, C. Subramanian and M. Gopinath: 'The Metallography of the Wear Processes of Coated Carbide Cutting Tools,' presented at the *Int. Conf. Metallurgical Coatings*, San Diego, April 1995.
7. S. Saunders and S. J. Bull: *5th Int. Conf. on Surface Modification Technologies*, Birmingham, Sept. 2–4, 1991.
8. A. Matthews, A. Leyland, R. Ronkainen, T. Nieminen, K. Holmberg, B. Matthes and E. Broszeit: *Standardisation of Test Procedures in Surface Engineering*, I.Mech.E., C412/038.
9. H. Heib: *Surface Engineering – Process and Applications*, K. N. Strafford, R. Smart, I. Sare and C. Subramanian, eds, Technomic, 1995, 217.
10. T. Sato, M. Tada, Y. C. Huang and H. Takei: *Thin Solid Films*, 1978, **54**, 61.
11. P. K. Datta, R. Hill, K. N. Strafford and L. Ward: *Surface Engineering Practice – Processes, Fundamentals and Applications in Corrosion and Wear*, K. N. Strafford, P. K. Datta and J. S. Gray, eds, Ellis Horwood, Chichester, UK, 1990, 89.
12. K. N. Strafford, C. Subramanian and T. P. Wilks: 'Surface Engineering – Vol. II Engineering Applications,' Royal Society of Chemistry, *Special Publication No. 127*, P. K. Datta and J. S. Gray, eds, 1993, 259.
13. S. J. Bull: This publication, 161.
14. A. Matthews and A. Leyland: This publication, 43.
15. H. M. Pollock: *ASM Handbook*, Volume 18 – *Friction, Lubrication and Wear Technology*, ASM, OH, 1992, 419.
16. G. R. Anstis, P. Chantikul, B. R. Lawn and D. B. Marshall: *J. American Ceramic Society*, 1981, **64** (9), 533.
17. D. T. Quinto and G. J. Wolfe: *Protective Coatings: Processing and Characterisation*, R. M. Yazici, ed., TMS, 1990.
18. H. Dagnall: *Exploring Surface Texture*, Rank Taylor Hobson, Leicester, UK, 1980.
19. M. Clarke: *Corrosion* Vol. 2 – *Corrosion Control*, L. I. Shreired., Newnes-Butterworths, 1976, Chapter 13.1, 13.37.
20. *Surface Engineering – Processes and Applications*, K. N. Strafford, R. Smart, I. Sare and C. Subramanian, eds, Technomic, 1995, XVII.
21. D. W. Borland: This publication, 115.
22. K. N. Strafford: 'Tribological Properties of Coatings – Expectations, Performance and the Design Dilemma,' *First Australian–U.S.A. Workshop on Critical Issues in High Performance Wear-resistant Films*, Wollongong, N.S.W., Australia, Feb 1–3, 1995. To appear in *Surface and Coatings Technology*, 1995.
23. K. N. Strafford, P. K. Datta and J. S. Gray: *Surface Engineering Practice – Processes, Fundamentals and Applications in Corrosion and Wear*, Ellis Horwood, Chichester, UK, 1990. Chapter 3.4.3, 398.
24. X. Chu, S. A. Barnett, M. S. Wong and W. D. Sproul: *Surface and Coatings Technology*, 1993, **57**, 13.
25. J. D. J. Ross, H. M. Pollock, J. C. Pivin and J. Takadoum: *Thin Solid Films*, 1987, **148**, 171.

26. M. McPhee, C. Subramanian, K. N. Strafford, T. P. Wilks and L. P. Ward: *Surface and Coatings Technology*, 1995, **71**, 215.
27. J. Audy, K. N. Strafford and C. Subramanian, to be published in *Surface and Coatings Technology.*
28. EC Brite Project BE – 3413 Design Methodology for Engineering Component Properties – Precept. March 1990.

Author Index

Subject Index

For Product Safety Concerns and Information please contact our EU representative GPSR@taylorandfrancis.com Taylor & Francis Verlag GmbH, Kaufingerstraße 24, 80331 München, Germany

Batch number: 10726412

Printed by Printforce, the Netherlands